ECOLOGICAL SPIRITUALITY: CROSS CULTURAL PERSPECTIVE

ECOLOGICAL SPIRITUALITY: CROSS CULTURAL PERSPECTIVE

Edited by

DAVIS GEORGE

J.G. VALAN ARASU

ANJALI D'SOUZA

NEELANJANA PATHAK

SERIALS PUBLICATIONS

NEW DELHI

First Published - 2007

ISBN : 978-81-8387-108-2

The opinions expressed in this edited volume are solely of the individual authors and the editors are in no way responsible for the views expressed by the article contributors.

Published by

SERIALS PUBLICATIONS
4830/24, Ansari Road, Darya Ganj
New Delhi-110002 (India)
Phone : 91-11-23245225
Fax : 91-11-23272135
E-mail: serials@satyam.net.in,
serialspublications@vsnl.net
Website : www.seriealspub.com

CONTENTS

PREFACE

From time immemorial, the quest for the truth has kept the humankind exploring and discovering the mysteries of life. The various discoveries at different levels have unraveled some great truths to encourage man to continue in this direction. Religion deals with the inner space and Science deals with the outer space. The Aloysian Explore and Worship Forum at St. Aloysius college has made combined effort in this exploration and is still moving on with great gusto and enthusiasm. It seeks to provide a consolidated platform to converge our thinking on these vital issues. Man needs to liberate himself from the superstitions, beliefs and dictatorship of science. This will enable our mankind to work for the betterment of the entire race. The objectives of "The Aloysian Explore and Worship Forum" have been to create awareness of relation between science and religion and to motivate individuals to explore the scientific aspects of religion so that proper understanding and the union of these two may preserve life to bring about unity and peace in this world.

In this pursuit 'The Aloysian Explore and Worship Forum' at St. Aloysius organized a two Days National Seminar on "Ecology and Religion: Ecological Spirituality in Cross Cultural Perspective" in collaboration with the Metanexus Institute, Philadelphia, USA, which has been promoting dialogue between science and religion at the international level. As eminent theologian Samuel Ryan says, "Earth system is being polluted and destroyed through wasteful, profligate and predatory practices by modern profit-oriented scientific-technological culture... The Earth's standing meaning and history as the Home of Life are under threat of death... Precious resources are diverted to destructive and poisonous purposes like arms race, stocking of nuclear weapons, star wars and similar insanities which can wreck and ruin, which have wrecked and ruined the Earth and life partially, totally, irreparably. Nuclear radioactive and other fatally toxic wastes are being dumped in the oceans or on islands and continents of the poor and the powerless. These, together with polluting affluences and emissions from factories, and the acid rains they spawn, are causing large-scale disease and death to plants, birds, beasts, fish, lakes, rivers and human beings. The atmosphere of most cities thick with toxic fumes has become a major menace to health and life... Chemical fertilizers and pesticides, poison and weaken the soil, paving the way to greater use of them and greater disasters for the health of the Earth and living things. Water is overused or wasted and underground water sources are being drained dry. Accumulation of Carbon dioxide in the atmosphere and the felling of tropical trees by the thousands to pamper to a consumerist and throw-away culture have led to rise in the Earth's temperature, drastic changes in climatic condition, and depletion of the Ozone layer. These are already threatening to reduce the Earth's luxuriant forests and fertile fields into deserts."

Delegates from various parts of the Country were invited to think and reflect on ecological spirituality from a cross cultural perspective. More than 180 participants and 160 papers were presented in this seminar and it is expected to initiate an era of renewed ecological consciousness and concern worldwide. The selected papers in this collection are by environmentalists, ecologists, researchers and academicians who share their concerns regarding ecological and spiritual responsibilities, particularly in the face of individual technological advancements during the new millennium. These papers are but a small sample of the many ecological concerns that have begun to tease educationalists to become active agents in the sustenance of a global climate which fosters a meeting ground between development and environmental preservation. It is time to realize the truth behind the weighty words of Dr. S. Radhakrishnan, "To be spiritual is not to reject reason, but to go beyond it".

With a broad spectrum of engaging themes from land vegetation patterns, relevance of herbal plants, man's onus towards his surroundings to ethical concerns like Euthanasia, the articles are sure to fill the gaps in our knowledge and generate new questions, regarding ecological and spiritual ties. Thus, for those who intend to bridge ecology and religion, this anthology may prove to be an indispensable source.

The editors would like to place on record their gratitude towards all the contributors for their sizzling ideas. We wish to thank Dr. Joydeep Dutta, Assistant Professor in Zoology at St. Aloysius College for his laborious hours of computer typing. We also express our deep sense of gratitude Fr. Thomas Mathew and to all Staff members of St. Aloysius College for their dedicated work and support in organizing the two days National Seminar. We also extend our heart felt thanks towards Serials Publications for bringing out this volume in an elegant form. We shall welcome the response and suggestions of our readers for our next editions.

It is hoped that in the ensuing years such discourses will sensitize us to the grim issue of continuous deterioration of our environment and the pressing need to evolve an international plan of action. Together, let us strive towards healthier living conditions, a greener planet and peaceful co-existence through enlightened and eco-conscious mind.

Davis George,
Valan Arasu,
Anjali D'Souza,
Neelanjana Pathak

CONTRIBUTORS

Davis George is Principal at St. Aloysius College (Autonomous), Jabalpur, M.P.

B.N. Behera is Head at Post Graduate Department of Botany, Rajendra College (Autonomous), Bolangir, Orissa.

P.K. Patra is Reader in Sanskrit, Rajendra College (Autonomous), Bolangir, Orissa.

G. Therasamma is Principal of St. Joseph's College of Education for Women, Guntur, A.P.

Mona Purohit is Senior Lecturer at Department of Post Graduate Studies and Research in Law, Rani Durgawati Vishwavidyalaya, Jabalpur, M.P.

J.G. Valan Arasu is Vice-Principal and Assistant Professor in Economics at St. Aloysius College (Autonomous), Jabalpur, M.P.

P.C. Aniyankunju, is Professor at B.A.M College, Thuruthicad, Kerala.

Yusuf Amin is Professor in Pharmacology at Aligarh Muslim University, Aligarh.

Jayasree K. is Lecturer at St. Joseph's College of Education for Women, Guntur.

Marcel Ekka, S.J. is Head and Assistant Professor in History at St. Aloysius Autonomous College, Jabalpur, M.P.

Gargi Mukhopadhyay is Lecturer at Department of Political Science, Bethune College, 181, Bidhan Sarani, Kolkata, West Bengal.

K. Narayanan is Principal at St. Mary's Syrian College, Brahmavar, Udupi, Karnataka.

J. Manoharan Thomas, Nazareth Margoschis College, Nazareth, Tamil Nadu.

Nandita Priyamvada Trivedi is Assistant Professor in Sociology at Government M.A.C. Autonomous College, Jabalpur.

Sudha Mehta is Assistant Professor in Psychology at Government M.K.B. College, Jabalpur.

Nirmala P. is Reader at Holy Cross College, Nagercoil, Kanyakumari, T.N.

Shobha Diwakar, Assistant Professor at C. P. Mahila College, Jabalpur.

Neelanjana Pathak, is Assistant Professor in English at St. Aloysius College (Autonomous), Jabalpur, M.P.

Shobha Menon is Senior Advocate at High Court of Madhya Pradesh, Jabalpur.

Jacintha is Vice Principal at J.M.J. College for Women, Tenali, Gurtur, A.P.

Rama Devi is Lecturer at J.M.J. College for Women, Tenali, Gurtur, Andhra Pradesh.

Lilly V.S. is Principal at Sacred Heart College, Chalakudy, Thrissur, Kerala.

Anjali D'Souza is Head and Assistant Professor in Chemistry at St. Aloysius Autonomous College, Jabalpur, M.P.

S.Thadedus Maria Ignatiammal is Reader at Department of Botany, Holy Cross College, Nagercoil, Tamil Nadu.

Aarti Kanojiya is Assistant Professor in Biochemistry at St. Aloysius Autonomous College, Jabalpur, M.P.

Priyanka Parmar is Assistant Professor in Electronics at St. Aloysius College (Autonomous), Jabalpur, M.P.

Deepa Pal is Assistant Professor in Electronics at St. Aloysius College Autonomous, Jabalpur, M.P.

Samuel Richmond Saxena is Assistant Chaplain in Department of Theology at Allahabad Agricultural Institute - Deemed University, U.P.

Vineeta Tiwari is Computer Operator at Computer Centre, R.D. University, Jabalpur, M.P. 482001.

Piku Dasgupta is Lecturer in Political Science at Bethune College, Bidhan Sarani, Kolkota.

Dipa Bardhan is Assistant Professor in Microbiology at St. Aloysius College (Autonomous), Jabalpur, M.P.

Vidhu G. Singh is A*d-hoc* Lecturer in Department of Business Administration at St. Johns College, Agra, U.P.

Asish S. Noel is A*d-hoc* Lecturer in Department of Accounts & Law at St. Johns College, Agra, U.P.

Pragyesh Agrawal is Head & Assistant Professor in Electronics at St. Aloysius College Autonomous, Jabalpur, M.P.

Varsha Gupta is Assistant Professor in Electronics at St. Aloysius College (Autonomous), Jabalpur, M.P.

J. Dutta is Assistant Professor in Zoology at St. Aloysius College (Autonomous), Jabalpur, M.P.

A. Nandurkar is Assistant Professor in Computer Science at St. Aloysius College (Autonomous), Jabalpur, M.P.

Mita Darbari is Head & Assistant Professor in Mathematics at St. Aloysius College (Autonomous), Jabalpur, M.P.

Elena Philip is Head in the Department of Commerce & Assistant Professor in Economics at St. Aloysius College (Autonomous), Jabalpur, M.P.

Peter Amalodoss is Professor in Philosophy at Christ Hall Seminary, Arul Anandar College, Karumathur, Madurai.

T. Kamala is Reader at Department of Biology at Holy Cross College, Nagercoil, Tamil Nadu.

Priti Gupta is Assistant Professor in Microbiology at St. Aloysius College (Autonomous), Jabalpur, M.P.

S.K. Khare is Head & Assistant Professor in Economics at St. Aloysius College (Autonomous), Jabalpur, M.P.

Jayanti Deshpande is Assistant Professor in Economics at St. Aloysius College (Autonomous), Jabalpur, M.P.

J.V. Zahir is Head & Assistant Professor in English at St. Aloysius College (Autonomous), Jabalpur, M.P.

ECOLOGICAL STEWARDSHIP: THE BIBLICAL PERSPECTIVE

Davis George

Ecological Challenges

The effects of ecological degradation surround us: the smog in our cities; chemicals in our water and on our food; eroded topsoil blowing in the wind; the loss of valuable wetlands; radioactive and toxic waste lacking adequate disposal sites, threats to the health of industrial and farm workers. The problems, however, reach far beyond our own neighborhoods and work-places. Our problems are the world's problems and burdens for generations to come. Poisoned water crosses borders freely. Acid rain pours on countries that do not create it. Greenhouse gases and chlorofluorocarbons affect the earth's atmosphere for many decades, regardless of where they are produced or used.

Opinions vary about the causes and the seriousness of environmental problems. Still, we can experience their effects in polluted air and water; in oil and wastes on our beaches; in the loss of farmland, wetlands, and forests; and in the decline of rivers and lakes. Scientists identify several other less visible but particularly urgent problems currently being debated by the scientific community, including depletion of the ozone layer, deforestation, the extinction of species, the generation and disposal of toxic and nuclear waste, and global warming. These important issues are being explored by scientists, and they require urgent attention and action. We are not scientists, but as responsible citizens of the world we call on experts, citizens, and policymakers to continue to explore the serious environmental, ethical, and human dimensions of these ecological challenges.

Ecological issues are also linked with other basic problems. As eminent scientist Dr. Thomas F. Malone reported, humanity faces problems in five interrelated fields: environment, energy, economics, equity, and ethics. To ensure the survival of a healthy planet, we must not only establish a sustainable economy but must also labor for justice both within and among nations. We must seek a society where economic life and environmental commitment work together to protect and to enhance life on this planet.

Ecology: A Common Patrimony

According to Pope John Paul II Ecology is our common patrimony. And the goods of the earth, which in the divine plan should be a common patrimony, often risk becoming the monopoly of a few who often spoil them and, sometimes, destroy them, thereby creating loss for all humanity. God has given the fruit of the earth to sustain the entire human family "without excluding or favoring anyone." The Second Vatican Council says "God destined the earth and all it contains for the use of every individual and all peoples."[1]

The Earth is a Gift to all Creatures

In the creation story we read in the Bible, "God looked at everything he had made, and he found it very good." (Gen 1:31) The heavens and the earth, the sun and the moon, the earth and the sea, fish and birds, animals and humans—all are good. The whole creation is called to bless the Lord. (Pro 8:2; Dan 3:74-81) The earth, the Bible reminds us, is a gift to all creatures, to "all living beings–all mortal creatures that are on earth." (Gen 9:16-17) Hence the covenant of Noah consisted of all creatures. (Gen 9:9-10) It is amazing to see how all living creatures are taken care of and protected by God himself. God's plan was that we live interconnected as we are interdependent.

Aquinas in Summa Theologica tells us that God produced many and diverse creatures. Hence the whole universe together participates in the divine goodness more perfectly, and represents it better than any single creature whatever. Respect for nature and respect for human life are inextricably related. "Respect for life, and above all for the dignity of the human person," Pope John Paul II has written, "extends also to the rest of creation."[2]

Alienation from Nature: Ecological Crisis

In the name of development man has been consistently alienating himself from nature. Exploitation and depletion of natural resources to satisfy man's insatiable lust and greed slowly made humanity more and more vulnerable to impoverished life and destruction. Human beings were made to be part of God's creation with an added responsibility and accountability to make this planet earth more productive and fruitful for all God's creation. Not paying heed to this sacred duty entrusted to him, man brought humanity to almost the verge of natural catastrophe.

Global Ecological Destruction: Consumption and Population

Consumption in developed nations remains the single greatest source of global ecological destruction. A child born in the United States, for example, puts a far heavier burden on the world's resources than one born in a poor developing country.

To deal with population problems the world has to focus on sustainable social and economic development. According to Gandhi, "There is enough in the world for man's need, but not enough for man's greed."

The Ecological Crisis: A Moral Problem

Faced with the widespread destruction of the environment, people everywhere are coming to understand that we cannot continue to use the goods of the earth as we have in the past. . . . New *ecological awareness* is beginning to emerge. . . . The ecological crisis is a moral issue.[3]

There is a growing awareness that world peace and prosperity is threatened not only by the arms race, regional conflicts and continued injustices among peoples and nations, but also by a lack of DUE RESPECT FOR NATURE, by the plundering of natural resources and by a progressive decline in the quality of life. The sense of precariousness and insecurity that such a situation engenders is a seedbed for collective selfishness, disregard for others and dishonesty. Faced with the widespread destruction of the environment, people everywhere are coming to understand that we cannot continue to use the goods of the earth as we have in the past. The public in general, as well as political leaders are concerned about this problem, and experts from a wide range of disciplines are studying its causes.

Pope John Paul II appreciated the great discoveries and technological advancements made by science. At the same time he expressed his concern over the indiscriminate application of advances in science and technology, which according to him have become a moral problem. As a result of this man is often oblivious of God's plan which is so evident in nature. Every time we discover something new, we discover how God has implanted his laws in the smallest of atom and the biggest of constellation. This should be a humbling experience which makes him wonder at the order in the universe and worship the God in nature. John Paul II expressed his concern over the growing lack of respect for life. God alone is the author of life and we need to learn to respect life and be grateful for the wonder of life in so many forms. Indiscriminate genetic manipulation can result in untold miseries. Human beings could be treated as any other animal and the bonding between person to person, parents and children, families and society may be lost. In the name of progress and scientific advancements, we should not manipulate and exploit human needs and human situation. Ethical values must be safeguarded to preserve human dignity.

Uncontrolled destruction of animal and plant life has brought about imbalance in ecology and this in turn has affected human beings. To add to this we have witnessed reckless exploitation of natural resources. We tend to forget that we are interdependent beings and man alone cannot survive on this planet earth. There is a growing awareness on this issue all around the world.

Authentic Development: Option for the Poor

The ecological problem is intimately connected with justice for the poor. Unrestrained economic development is not the answer to improving the lives of the poor. Material growth alone will not constitute a model of development. A "*mere accumulation* of goods and services, even for the benefit of the majority," as Pope John Paul II has said, "is not enough for the realization of human happiness."[4] He has also warned that in a desire "to have and to enjoy rather than to be and to grow," humanity "consumes the resources of the earth, subjecting it without restraint as if it did not have its own requisites and God-given purposes."

It must also be said that a proper ecological balance will not be found without directly addressing the structural forms of poverty that exist throughout the world. Rural poverty and unjust land distribution in many countries, for example, have led to subsistence farming and to the exhaustion of the soil. Once their land yields no more, many farmers move on to clear new land, thus accelerating uncontrolled deforestation, or they settle in urban centers which lack the infrastructure to receive them. Likewise, some heavily indebted countries are destroying their natural heritage, at the price of irreparable ecological imbalances, in order to develop new products for export. In the face of such situations it would be wrong to assign the responsibility to the poor alone for the negative environmental consequences of their actions. Rather, the poor, to whom the earth is entrusted no less than to others, must be enabled to find a way out of their poverty. This will require a courageous reform of structures, as well as new ways of relating among peoples and States.

The Biblical Vision of God's Good Earth

We read in the Bible, "So God created man in his own image, in the image of God he created him; male and female he created them. And God blessed them, and God said to them, "Be fruitful and multiply, and fill the earth and subdue it; and have dominion over the fish of the sea and over the birds of the air and over every living thing that moves upon the earth." (Gen 1:27-28)

Yet, man's lordship is not "absolute, but ministerial: it is a real reflection of the unique and infinite lordship of God. Hence, man must exercise it with wisdom and love, sharing in the boundless wisdom and love of God."[5] In Biblical language, "to name" creatures (Genesis 2:19-20) is the sign of this mission of knowledge and transformation of created reality. It is not the mission of an absolute and insensitive master, but of a minister of the Kingdom of God, called to continue the work of the Creator, a work of life and peace. His responsibility, defined in the Book of Wisdom, is to govern "the world in holiness and justice" (Wisdom 9:3; Wisdom 13:5; Romans 1:20). The Book of Wisdom, echoed by Paul, celebrates this presence of God in the universe. This is what the Jewish tradition of the Hasidim also sings "You are wherever I go! You are wherever I stop... wherever I turn, wherever I admire, only You, again You, always You."[6]

Bible and Ecology: Splendours of Creation

The book of Genesis teaches us that the Lord God formed us "out of the dust of the ground" (Gen 2:7; 3:19). Psalm 139 thanks God for fashioning us fearfully and wonderfully "in secret", "in the depths of the earth". The Psalms delight at and are full of awe over the mystery of our intimacy with the earth, our intimacy with "fire and hail, snow and mist", "mountains and all hills", "sea monsters and all depths" (Ps 148). Psalm 104, one of the most lyrical praises, sings the glory of God "robed in light as with a cloak", who "spread out the heavens like a tent cloth" and "made the moon to mark the seasons".

The Bible shows nature's link with God who created it, blessed it, and shows himself through it. He appears in fire, in wind, and in water. God also uses nature to bring humans closer to him and to punish them when they go astray. Everything in the world, therefore, remains sacred since it is linked with God and leads to him. Various texts in the Psalms (Ps 19:1-7; 98:7-9; 104:1-5, 13-25; 148:3-13) show that all things on earth are seen as God's handiwork which bring him honour and praise by their very existence.

The prophet Daniel in a canticle calls on all the "works of the Lord" to bless him: "Let the earth bless the Lord; praise and exalt him above all forever. Mountains and hills, bless the Lord, everything growing from the earth bless the Lord" (Dan 3:74-76). The last chapters of the Book of Job call upon the animals, nature, birds, etc., and praise God for their presence. Chapter 12 urges humans to learn humbly from the earth: "But ask the beasts, and they will teach you; the birds of the air, and they will tell you; or the plants of the earth, and they will teach you" (Job 12:7, 8).15 The Bible is concerned with salvation or life-giving blessings not only in the afterlife but also within this world and within present history, individual and collective. It envisions a new world and a new history. Its salvific concern embraces nature, that is, the earth, air, trees, seas and birds.

The cosmos is God's 'womb', as it were. The intimate relationship between God and the cosmos explodes with seminal energy that generates and regenerates life. God, as it were, energises the cosmos and the cosmos in return dances with the creator.

In Jesus' teaching, one can see his ecological concern in his language. He used ordinary creatures such as birds, lilies, grass, etc., to help to put his message of concern for the world across. He also shared his experience of a loving God dynamically present in the world. He is encouraging his listeners to have eyes that see and ears that hear the movement of God in the world. Jesus was passing on to his listeners what he had discovered about God's reign in the natural things around him.

The miracles of Jesus (37 of them in the Synoptic Gospels and seven in John) form a major section of the Gospels and reveal Jesus' concern for the world as such. Through the miracles Jesus destroys the "domination" of Satan over the created realities and establishes the "dominion" of God which is liberating. In this sense all the miracles have ecological resonance. The nature miracles (Mk 4:35-41; 6:45-62, etc.) invite us to trust in the absolute power of God in the midst of ecological disasters. The feeding miracles (Mk 6:30-44; 8:1-10) tell us about the abundant resources of nature, which provide us with food and drink, and which need to be evenly distributed according to the needs of the people. The miracles of exorcism (Mk 5:2-20; Lk 4:35-41, etc.) reveal that cosmic ecological harmony is on the agenda of God who directs the forces of ecocide. The healing miracles (Mk 5:25-34, etc.) call us to be God's stewards in the restoration of the disfigured images of God in creation, especially, human beings. The resuscitation miracles (Mk 5:21-21, 35-43, etc.) challenge us not to be silent spectators of the world-wide ecological holocaust that is taking place, but to be active agents in the creation of "a new heaven and a new earth" (Rev 21:1-4).

A serious reflection on the life-events of Jesus Christ, his teachings and his miracles from an ecological point of view is very inspiring. Today, if one reads the Gospel from an ecological perspective one can see Jesus of the Gospel as an 'Ecologist.'

Scientists Learn from Creation

Louis Agassiz, perhaps the greatest natural scientist of the nineteenth century, declared that it is the job of prophets and scientists alike to proclaim the glories of God and he spent his life as a scientist doing exactly that. As Ralph Waldo Emerson stated, "The true doctrine of omnipresence is that God reappears with all his parts in every moss and cobweb."[7] In our study of natural objects we are approaching the thoughts of the Creator, reading his conceptions, interpreting a system that is His and not ours."[8] "Facts are the words of God, and we may heap them together endlessly, but they will teach us little or nothing till we place them in their true relations, and recognize the thought that binds them together."[9]

Today there is another interesting trend. It is that the number of inventions based on copying nature is now beginning to be systematically exploited. In so doing, one need not even bring up the argument over whether "nature" refers to the handiwork of God or millions of years of mindless evolution; all that matters is that nature is incredibly successful at solving problems with which we have struggled for years.

This trend began by noticing that many inventions were discovered by observing how "nature" had solved problems. Inventors spent centuries trying to invent the airplane after watching birds fly. The book Serendipity: Accidental Discoveries in Science chronicles several of these observations which led to inventions. It also includes some discoveries that really appear to have been accidents, but many came from simply noticing the invention already working in nature, and using scientific inquiry to discover just how they work. For example, the color purple is associated with royalty partly because the natural dye Tyrian purple could only be extracted from small mollusks in the Mediterranean Sea. It was very expensive because it took 9,000 of them to produce a gram of dye. The synthesis of this color by William Perkin led to the birth of the synthetic dye industry. Certain peptides which are highly effective in fighting a variety of bacteria were discovered when it was observed that some African frogs would heal perfectly in murky water filled with lethal bacteria. The list goes on and on.[10]

The present scenario fosters the trend of a more systematic imitation of nature. The word "biomimicry" has been coined to refer to the idea of purposely copying nature to discover new inventions. The author of a book with that title sees this emerging field as the result of centuries of trying to fight nature as gradually succumbing to a trend to acknowledge nature's ways as best. She points out that not only has nature already invented everything we have but, it has many more inventions whose workings still evade us.

We realize that all our inventions have already appeared in nature in a more elegant form and at a lot less cost to the planet. Our most clever architectural struts and beams are already featured in lily pads and bamboo stems. Our central heating and air conditioning are bested by the termite tower's steady 86 degrees F. Our most stealthy radar is hard of hearing compared to the bat's multi-frequency transmission. And our new 'smart materials' can't hold a candle to the dolphin's skin or to the butterfly's proboscis. Even the wheel, which we always took to be a uniquely human creation, has been found in the tiny rotary motor that propels the flagellum of the world's most ancient bacteria.

Humbling also are the hordes of organisms casually performing feats we can only dream about. Bioluminescent algae splash chemicals together to light their body lanterns. Arctic fish and frogs freeze solid and then spring to life, having protected their organs from ice damage. Black bears hibernate all winter without poisoning themselves on their urea, while their polar cousins stay active, with a coat of transparent hollow hairs covering their skins like the panes of a greenhouse. Chameleons and cuttlefish hide without moving, changing the pattern of their skin to instantly blend with their surroundings. Bees, turtles, and birds navigate without maps, while whales and penguins dive without scuba gear. How do they do it? How do dragonflies outmaneuver our best helicopters? How do hummingbirds cross the Gulf of Mexico on less than one tenth of an ounce of fuel? How do ants carry the equivalent of hundreds of pounds in a dead heat through the jungle?

No wonder that these marvelous creations inspire awe and reverence; they are the work of the Almighty. When we look on any or the least of these, we are looking at God moving in his majesty and power. The new millennium promises to provide many new and wonderful inventions as scientists recognize the hand of God in nature and begin to understand the principles behind so many inventions which are found everywhere in His creations.

Estrangement of Humans from Nature: Ecological Conversion

In the Book of Genesis, where we find God's first self-revelation to humanity (Gen 1-3), there is a recurring refrain: "AND GOD SAW IT WAS GOOD". After creating the heavens, the sea, the earth and all it contains, God created man and woman. At this point the refrain changes markedly: "And God saw everything he had made, and behold, IT WAS VERY GOOD" (Gen 1:31). God entrusted the whole of creation to the man and woman, and only then as we read could he rest "from all his work" (Gen 2:3). Adam and Eve's call to share in the unfolding of God's plan of creation brought into play those abilities and gifts which distinguish human beings from all other creatures. At the same time, their call established a fixed relationship between mankind and the rest of creation. Made in the image and likeness of God, Adam and Eve were to have exercised their dominion over the earth (Gen 1:28) with wisdom and love. Instead, they destroyed the existing harmony BY DELIBERATELY GOING AGAINST THE CREATOR'S PLAN, that is, by choosing to sin. This resulted not only in man's alienation from himself, in death and fratricide, but also in the earth's "rebellion" against Him (Gen 3:17-19; 4:12).

In the Bible's account of Noah, the world's new beginning was marked by the estrangement of humans from nature. Hosea, for example, cries out:

> There is no fidelity, no mercy,
> no knowledge of God in the land.
> False swearing, lying, murder, stealing
> and adultery!
> in their lawlessness, bloodshed
> follows bloodshed.
> Therefore, the land mourns,
> and everything that dwells in it
> languishes:
> The beasts of the field,
> the birds of the air,
> and even the fish of the sea perish (Hos 4:1b-3).

The idea of social justice is inextricably linked with ecology in the Scriptures. In passage after passage, environmental degradation and social injustice go hand in hand. Indeed, the first instance of "pollution" in the Bible occurs when Cain slays Abel and his blood falls on the ground, rendering it fallow. According to Genesis, after the murder, when Cain asks, "Am I my brother's keeper?" the Lord replies, "Your brother's blood calls out to me from the ground. What have you done?" God then tells Cain that his brother's blood has defiled the ground and that as a result, "no longer will it yield crops for you, even if you pray."

In the biblical vision, therefore, injustice results in suffering for all creation. To curb the abuse of the land and of fellow humans, ancient Israel set out legal protections aimed at restoring the original balance between land and people (Lev 25). Every seventh year, the land and people was to rest; nature would be restored by human restraint. And every seventh day, the Sabbath rest gave relief from unremitting toil to workers and beasts alike.

Pope John Paul II emphasized the need for personal conversion. "As individuals, as institutions, as a people, we need a change of heart to preserve and protect the planet for our children and for generations yet unborn."[11] We need to have a paradigm shift-from a culture of consumption to a culture of conserving; from depleting to replenishing.

Environmental Stewardship: God's Stewards and Co-Creators Stewardship: Protecting the Environment for Future Generations

Stewardship is defined in this case as the ability to exercise moral responsibility to care for the environment. It implies that we must both care for creation according to standards that are not of our own making, and at the same time be resourceful in finding ways to make the earth flourish. In Genesis, God said "till it and keep it", (Gen 2:15) and this should be understood not as dominion over the whole world, but as the 'stewardship' of human beings over the creatures. We must have a relationship of mutuality with other creatures and we must empathize and participate with, delight in, and accompany the creatures to bring about a communion of all sections of creation whose head is God himself. It is awesome that the creator of this universe in his wisdom entrusted his own creation to human beings so that they may take care of it and make it productive and fruitful for the benefit of the entire of creation. He did not visualize that humans would exploit the creation for his selfish ends. Yet, God alone is sovereign over the whole earth. "The LORD'S are the earth and its fullness; the world and those who dwell in it" (Ps 24:1). We are not free, therefore, to use created things capriciously. Humanity's arrogance and acquisitiveness, however, led time and again to our growing alienation from nature (Gen 3:4; 6:9, 11)

Theological and Ethical Foundations of Stewardship

God, the Creator of all things, rules over all and deserves our worship and adoration (Ps. 103:19–22). The earth, and, with it, all the cosmos, reveals its Creator's wisdom and goodness (Ps. 19:1–6) and is sustained and governed by his power and loving-kindness (Ps. 102:25–27; Ps. 104; Col. 1:17; Heb. 1:3, 10–12). Men and women were created in the image of God, given a privileged place among creatures, and commanded to exercise stewardship over the earth (Gen. 1:26–28; Ps. 8:5). Our stewardship under God implies that we are morally accountable to him for treating creation in a manner that best serves the objectives of the kingdom of God. However, both moral accountability and dominion over the earth depend on the freedom to choose. The exercise of these virtues and this calling, therefore, require that we act in an arena of considerable freedom-not unrestricted license, but freedom exercised within the boundaries of God's moral law revealed in Scripture and in the human conscience (Exod. 20:1–17; Deut. 5:6–21; Rom. 2:14–15). These facts are not vitiated by the fact that humankind fell into sin (Gen. 3). Rather, our sinfulness has brought God's responses, first in judgment, subjecting humankind to death and separation from God (Gen. 2:17; 3:22–24; Rom. 5:12–14; 6:23) and subjecting creation to the curse of futility and corruption (Gen. 3:17–19; Rom. 8:20–21); and then in restoration, through Christ's atoning, redeeming death for his people, reconciling them to God (Rom. 5:10–11, 15–21; 2 Cor. 5:17–21; Eph. 2:14–17; Col. 1:19–22), and through his wider work of delivering the earthly creation from its bondage to corruption (Rom. 8:19–23). Indeed, Christ even involves fallen humans in this work of restoring creation (Rom. 8:21). As Francis Bacon wrote in *Novum Organum Scientiarum* (New Method of Science), that man by the 'Fall' fell at the same time from his state of innocence and from his dominion over creation. Both these losses, however, can even in this life be in some parts repaired; the former by religion and faith, the latter by the arts and sciences. Sin, then, makes it difficult for humans to exercise godly stewardship, nonetheless the work of Christ in, on, and through his people and the creation makes it largely possible.

When He created the world, God set aside a unique place, the Garden of Eden, and placed in it the first man, Adam (Gen. 2:8–15). God instructed Adam to cultivate and guard the Garden (Gen. 2:15)-to enhance its already great fruitfulness and to protect it against the encroachment of the surrounding wilderness that made up the rest of the earth. Having also created the first woman and having joined her to Adam (Gen. 2:18–25), God commanded them and their descendants to multiply, to spread out beyond the boundaries of the Garden of Eden, and to fill, subdue, and rule the whole earth and everything in it (Gen. 1:26, 28). Both by endowing them with his image and by placing them in authority over the earth, God gave men and women superiority and priority over all other earthly creatures. This implies that proper environmental stewardship, while it seeks to harmonize the fulfillment of the needs of all creatures, nonetheless puts human needs above non-human needs when the two are in conflict.

Some environmentalists reject this vision as "anthropocentric" or "speciesist," and instead promote a "biocentric" alternative. But the alternative, however attractively humble it might sound, is really untenable. People, alone among creatures on earth, have both the rationality and the moral capacity to exercise stewardship, to be accountable for their choices, to take responsibility for caring not only for themselves but also for other creatures. To reject human stewardship is to embrace, by default, no stewardship. The only proper alternative to selfish anthropocentrism is not biocentrism but Theo centrism: a vision of earth care with God and his perfect moral law at the center and human beings acting as his accountable stewards.

Authentic Development Based on Justice

How are we to fulfill God's call to be stewards of creation in an age when we may have the capacity to alter that creation significantly and perhaps irrevocably? How can we as a "family of nations" exercise stewardship in a way that respects and protects the integrity of God's creation and provide for the common good? For this we need to focus on economic and social progress based on justice. Sustainable development can happen, only when we focus on justice; justice not only for the rich, but also for the poor and marginalized; even to all other plant and animal kingdoms. In the name of development we have often destroyed the mother earth on whom we depend for survival. Exploitation of the environment has resulted in ecological imbalance and poses great threat to future of the planet earth. The common good calls us to extend our concerns to future generations. Climate change poses the question "What does our generation owe to generations yet unborn?" As Pope John Paul II has written, "there is an order in the universe which must be respected, and . . . the human person, endowed with the capability of choosing freely, has a grave responsibility to preserve this order for the well-being of future generations."[12]

Passing along the problem of global climate change to future generations as a result of our delay, indecision, or self-interest would be easy. But we simply cannot leave this problem for the children of tomorrow. As stewards of their heritage, we have an obligation to respect their dignity and to pass on their natural inheritance, so that their lives are protected and, if possible, made better than our own.

A more responsible approach to population issues is the promotion of "authentic development," which represents a balanced view of human progress and includes respect for nature, respect for order in the universe and social well-being.

Interdependence to Solidarity and Moral Responsibility

"The ecological crisis," Pope John Paul II has written, "reveals the urgent moral need for a new solidarity, especially in relations between the developing nations and those that are highly industrialized".[13] The earth's atmosphere encompasses all people, creatures, and habitats. The melting of ice sheets and glaciers, the destruction of rain forests, and the pollution of water in one place can have environmental impacts elsewhere. As Pope John Paul II has said, "We cannot interfere in one area of the ecosystem without paying due attention both to the consequences of such interference in other areas and to the well being of future generations."[14] Responses to global climate change should reflect our interdependence and common responsibility for the future of our planet. Individual nations must measure their own self-interest against the greater common good and contribute equitably to global solutions.

Pope John Paul II has said that interdependence must be transformed into solidarity. Surmounting every type of imperialism and determination to preserve their own hegemony, the stronger and richer nations must have a sense of moral responsibility for the other nations, so that a real international system may be established which will rest on the foundation of the equality of all peoples and on the necessary respect for their legitimate differences."[15] Whether we like it or not, we have all been born on this earth as part of one great family. Rich or poor, educated or uneducated, belonging to one nation, religion, ideology or another, ultimately each of us is just a human being just like everyone else. We all desire happiness and do not want suffering. Furthermore, each of us has the same right to pursue happiness and avoid suffering. When we recognize that all beings are equal in this respect, we automatically feel empathy and closeness for them. Out of this, in turn, comes a genuine sense of universal responsibility: the wish to actively help others overcome their problems.

Nowadays, significant events in one part of the world eventually affect the entire planet. Therefore, we have to treat each major local problem as a global concern from the moment it begins. We can no longer invoke the national, racial or ideological barriers that separate us without destructive repercussions. Tenzin Gyatso Dalai Lama of Tibet said that in the face of such global problems as the greenhouse effect and depletion of the ozone layer, individual organizations and single nations are helpless. Unless we all work together, no solution can be found. Our mother earth is teaching us a lesson in universal responsibility. Good wishes alone are not enough; we have to assume responsibility.

Conclusion: Need of the Hour-Ecological Balance

We are amazed to see the marvels of science and technology and at the same time saddened to see human starvation in some parts of the world and extinction of other life forms. Where exploration of outer space is taking place paradoxically the earth's own ocean, seas and freshwater areas grow increasingly polluted, and their life forms are still largely unknown or misunderstood...Many of the earth's habitats, animals, plants, insects, and even microorganisms that we know as rare may not be known at all by future generations. We have the capability and the responsibility to take proactive steps to preserve the ecological balance. The whole universe is God's dwelling. Earth, a very small, uniquely blessed corner of the universe, gifted with unique natural blessings, is humanity's home, and humans are never so much at home as when God dwells with them. In the beginning, the first man and woman walked with God in the cool of the day. Throughout history, people have continued to meet the Creator on mountaintops, in vast deserts, and alongside waterfalls and gently flowing springs. In storms and earthquakes, they found expressions of divine power. In the cycle of the seasons and the courses of the stars, they have discerned signs of God's fidelity and wisdom. We still share, though dimly, in that sense of God's presence in nature.

For many people, the environmental movement has reawakened appreciation of the truth that, through the created gifts of nature, men and women encounter their Creator. The Christian vision of a sacramental universe–a world that discloses the Creator's presence by visible and tangible signs–can contribute to making the earth a home for the human family once again. Pope John Paul II has called for Christians to respect and protect the environment, so that through nature people can "contemplate the mystery of the greatness and love of God."

Reverence for the Creator present and active in nature, moreover, may serve as a ground for ecological responsibility. For the very plants and animals, mountains and oceans, which in their loveliness and sublimity lift our minds to God, by their fragility and perishing likewise cry out, "We have not made ourselves." God brings them into being and sustains them in existence. It is to the Creator of the universe, then, that we are accountable for what we do or fail to do to preserve and care for the earth and all its creatures. For "the LORD'S are the earth and its fullness; the world and those who dwell in it" (Ps 24:1). Dwelling in the presence of God, we begin to experience ourselves as part of creation, as stewards within it, not separate from it. As faithful stewards, fullness of life comes from living responsibly within God's creation.

Stewardship implies that we must both care for creation according to standards that are not of our own making and at the same time be resourceful in finding ways to make the earth flourish. It is a difficult balance, requiring both a sense of limits and a spirit of experimentation. Even as we rejoice in the earth's goodness and in the beauty of nature, stewardship places upon us the responsibility of the well-being of all God's creatures.

Education in ecological responsibility is urgent if our "paradise lost" has to be regained. The children have to be taught to respect their neighbors and to love nature. Lyndon B. Johnson, the 36th President of the United States, said that if future generations are to remember us with gratitude rather than contempt, we must leave them more than the miracles of technology. We must leave them a glimpse of the world as it was in the beginning, not just after we got through with it.

Notes

1. Second Vatican Council, Pastoral Constitution on the Church in the Modern World (Gaudium et Spes), no. 69, in Austin Flannery, ed., Vatican Council (Gaudium et Spes, 69).
2. John Paul II, The Ecological Crisis: A Common Responsibility (Washington, D.C.: United States Conference of Catholic Bishops, 1990), no. 7.
3. Ibid., nos. 1, 15.
4. John Paul II, On Social Concern (Sollicitudo Rei Socialis) (Washington, D.C.: United States Conference of Catholic Bishops, 1988), no. 28.
5. John Paul II, Evangelium Vitae," NewYork, Random House, 1995, no. 52.
6. M. Buber," I Racconti dei Chassidium," Milan 1079, p. 256.
7. T Royston M. Roberts, Serendipity: Accidental Discoveries in Science, Wiley & Sons, New York, 1989.
8. The Agassiz, Louis, *Methods of Study in Natural History*, Boston: Ticknor and Fields, 1863, p. 14.
9. Agassiz, Louis, "Evolution and Permanence Type" reprinted in *The Intelligence of Agassiz* by Guy Davenport, Westport, Conn., Greenwood Press, 1983, p. 231.
10. Benyus, Janine M., *Biomimicry*, William Morrow, New York, 1997, pp. 6-7.
11. John Paul II, The Ecological Crisis: A Common Responsibility (Washington, D.C.: United States Conference of Catholic Bishops, 1990), no. 6.
12. John Paul II, "The Exploitation of the Environment Threatens the Entire Human Race," address to the Vatican symposium on the environment (1990), in Ecology and Faith: The Writings of Pope John Paul II, ed. Sr. Ancilla Dent, OSB (Berkhamsted, England: Arthur James, 1997), 12.
13. John Paul II, The Ecological Crisis: A Common Responsibility (Washington, D.C.: United States Conference of Catholic Bishops, 1990), no. 10.
14. Ibid., no. 6.
15. See also treatment of this topic in Stewardship: A Disciple's Response (Washington, D.C.: United States Conference of Catholic Bishops, 1993), 39.

References

Second Vatican Council, Pastoral Constitution on the Church in the Modern World (Gaudium et Spes), No. 69, in Austin Flannery, ed., Vatican Council (Gaudium et Spes, 69).

John Paul II, The Ecological Crisis: A Common Responsibility (Washington, D.C.: United States Conference of Catholic Bishops, 1990).

John Paul II, On Social Concern (Sollicitudo Rei Socialis) (Washington, D.C.: United States Conference of Catholic Bishops, 1988).

John Paul II, Evangelium Vitae," NewYork, Random House, 1995.

M. Buber," I Racconti dei Chassidium," Milan 1079.

T. Royston M. Roberts, Serendipity: Accidental Discoveries in Science, Wiley & Sons, New York, 1989.

The Agassiz, Louis, *Methods of Study in Natural History*, Boston: Ticknor and Fields, 1863.

Agassiz, Louis, "Evolution and Permanence Type" reprinted in *The Intelligence of Agassiz* by Guy Davenport, Westport, Conn., Greenwood Press, 1983.

Benyus, Janine M., *Biomimicry*, William Morrow, New York, 1997.

John Paul II, "The Exploitation of the Environment Threatens the Entire Human Race," address to the Vatican symposium on the environment (1990), in Ecology and Faith: The Writings of Pope John Paul II, ed. Sr. Ancilla Dent, OSB (Berkhamsted, England: Arthur James, 1997).

Wallace, Mark I. Fragments of the Spirit: Nature, Violence, and the Renewal of Creation. New York: Continuum, 1996.

Williams, George H. "Christian Attitudes toward Nature." Parts 1 and 2. Christian Scholar's Review 2, No. 1 (fall 1971): 3–35; No. 2 (spring 1972): 112–26. Wilderness and Paradise in Christian Thought. New York: Harper and Brothers, 1962.

Zizioulas, John. "Preserving God's Creation: Three Lectures on Theology and Ecology." Parts 1–3. King's Theological Review 12 (spring 1989): 1–5; 12 (autumn 1989): 41–45; 13 (spring 1990): 1–5.

John Paul II, On the Hundredth Anniversary of Rerum Novarum (Centesimus Annus) (Washington, D.C.: United States Conference of Catholic Bishops, 1991).

John Paul II, "International Solidarity Needed to Safeguard Environment," Address by the Holy Father to the European Bureau for the Environment, L'Osservatore Romano (June 26, 1996).

REDISCOVERY OF TRADITIONAL PERSPECTIVE WILL PROVIDE THE TRUE PHILOSOPHY AND SOUND SCIENCE NEEDED BY ECOLOGY

Yusuf Amin

Ecology suddenly shot from obscurity to tremendous popularity towards the end of the last century. However, the degree of its hold is matched by the challenges that beset it, both as a science and as an agenda. It is most complex, least amenable to scientific study and suspected to provide only partial solutions to the ecological crisis. Its conclusions demand nearly all of a Man, the renunciation of the deepest attitudes of the modern man and adoption of ideas and practices that have been all but forgotten. Religion is perceived to help on both counts: arriving at prescriptions and putting them to practice.

Reviewers of Ecology admit that its models are crude and have a doubtful predictive value. For instance: Vegetarianism is held out as ecologically desirable, almost essential. But very limited implementations of the idea have shown its harmfulness. The ban on the hunting of Nilgais in North India have produced such an increase in their numbers that they are causing large scale damage to crops and causing frequent accidents on highways. Further, a faithful adoption of the principle is quite likely to cause malnutrition, at least to the less aware and resourceful.

The logically most full-blown versions of Ecology, considered close to the religious perspective, for instance as reviewed in Capra's "Web of Life", are beset with fundamental problems too. They consider the vision of Man's amalgamation with Nature to be enough for altruistic behaviour demanded by ecological prescriptions. But this is fallacious as selfish behaviour is not cognitive, arising from perception of a separated ego but affective and born out of the pleasure of personal consumption and possession. Secondly, they base all phenomena, particularly the supra-organism entities such as populations, ecosystems and the biosphere on 'organising principles' rather than on supra-material substance, a position which is intellectually incomprehensible, to say the least.

The extreme limitations of Ecology are evident at a very obvious level too. It deals with a very huge number of variables with much more numerous interconnections of diverse types. A very small fraction of them are known. To say something beyond the most basic generalities on their basis can be not only misleading but highly dangerous too. The analogy of biomedical Medicine amply illuminates the matter. Based as it is only on a very small fraction of disjointed description of human molecular biology and biochemistry, it often does more harm than good. The object of Ecology, i.e. the Biosphere is much vaster than Man, so a purely materialistic scientific approach can be of more devastating outcome.

Another basic shortcoming of modern ecology is the absence of perspective and intellectual tools that are commensurate with the centrality of Man in this matter. It recognizes the crucial role of Man, both his preciousness as well as dangerousness, in Ecology. But this is dealt merely by using Anthropology, probably one of the weakest of the generally imprecise modern Social Sciences, that does not even enjoy independent departments in present day Universities.

The cognitive problems are aggravated by the deep presumption that Science and its application to the Biosphere, namely, Ecology is the only means of getting knowledge of the principles for Man's harmonization with it. It is indeed surprising that a method of knowing that itself knows its extreme limitations to know its objects (this realization being only the tip of the iceberg of its real epistemic insufficiency) holds itself to be the only method of harmonizing with Nature with such tragic certainty. Man has lived in harmony with Nature from his advent on the Earth up to the Renaissance without the help of modern Ecology not because he was not powerful enough, unarmed by Baconian Science, but because he enjoyed and followed illuminative guidance, exactly the thing of which ecology is so ignorant as to not even deny it explicitly. In fact such a Science could develop only when Man became veiled from this knowledge.

We so far have discussed the descriptive and prescriptive problems regarding Ecology. The implementation of its prescriptions, entailing immense personal and national sacrifices as they do, have very little chance of being made, motivated only by a fear of distant adversities, there also being the further assurance of postponing them by arm-twisting weaker nations to shun even minimal 'development' and thus keeping the global side-effects within acceptable limits. For instance, Ecology demands cycling, organic farming, making do with solar heating, etc, not to talk of more basic revolutions like the replacement of competition by cooperation, domination by partnership, invention of wants by reduction of needs and slowing the overall pace of 'Development'. Can even Third World elites, not to speak of Johnny heads-in-air American and European ruling classes, be expected to follow these guidelines?

May be it was the motivational difficulties that made people look to Religion for succour. The United Nations Environment Programme (UNEP) has established an Interfaith Partnership for the Environment (IPE) being a global formalization of this feeling. But, as expected Religion is approached in a typical Post-Renaissance instrumental manner, only as an aspect of human culture, enjoying no privilege, much less primacy, in knowing truth or deserving obedience. It is a token of modern simple mindedness to aim to seek motivation from an intellectually and ethically subservient religion whose power arises exactly from the acceptance of its sovereignty to teach and to command. This does not mean denial of human experience, but its acceptance on word of, and under qualifications by Religion, and not as a parallel, much less primary entity.

Four factors are essential for ensuring harmony between Man and Nature: a Philosophy of Nature and Cosmology, a Philosophy of Man and Psychology, a sound Science and a sound Ecology, and all of them are provided by Religion through the mediation of Metaphysics and Philosophy. This is the great role of Religion in Ecology and not merely to egg on the feeble-minded masses that otherwise cannot understand and practice the scientific Ecology.

The modern world and its ecology lack not only a Philosophy of Nature and Man but also Philosophy as such. What passes under its name are generalizations, clarifications and validations of the implications of modern sciences, literature, art, etc. True Philosophy comprises of a multifarious Reason: dialectical, analogical, etc. applied to the fruits of illuminative intellection, obtained and nurtured in the bosom of Revelation vouchsafed to, and efficacious in, only the protective shell of a heavenly, traditional religion.

The Philosophy of Nature thus arising and subsisting in the Islamic Universe, which inevitably has correspondences in all other religious universes-Christian, Judaic, Hindu, Buddhist, Taoist, Shintoist–envisions it as a Theophany whose primary purpose is to reveal God, including His mercy, manifested among other things by Nature being made amenable to the fulfillment of Man's worldly needs and desires. So, Nature is living, sacred and primarily a means of contemplation also made amenable to Man for the fulfillment of his valid need in moderation, which has to be achieved by cooperating with it. Such an I–Thou relation not only provides the perspective for a sound Cosmology and Science but is also in itself a very effective means of directly ensuring harmony between Man and Nature.

This Philosophy is accompanied by ethical and legal implications, which are more explicitly obtained from Revelation. The Canonical prescriptions and their juristic applications are of greater importance in Ecology than other aspects of life because they are less easily discoverable by Science on account of the greater heterogeneity and complexity of the Biosphere.

It must be noted that Scripture and its Rational-Canonical-Theological understanding is not enough for arriving at a Philosophy. It needs the mediation of Metaphysics arising from Revelation and Intuitive Intellection and systematization. In fact Metaphysics is indispensable for a comprehensive Theology too, as well as for Cosmology and Philological, Social and Natural Sciences.

A Philosophy or Theology is not enough for satisfactorily understanding and using Nature, a Cosmology is also needed. A Cosmology is a substantial description of all grades of formal manifestation, including the material level of Nature. Islamic Cosmologies describe the material, subtle, angelic and celestial levels and the Absolute. The subtle-material level is comprised of four elements Fire, Air, Water and Earth. All physical phenomena including the body of man are made by the amalgamation of the elements. There are very similar to Hindu, Empedoclean, Neo-Platonic Cosmologies, etc. They provide the substances unifying fragmented material phenomena which Capra and all Post-Renaissance, modern, materialistic men so unconvincingly and confusingly replace by 'organising principles'. Further, they allow the pursuance of the study of material objects to any of their higher level, allowing not only the actualization of their contemplative function but making possible holistic study of phenomena. Tib-e-Unani–as well as all other Traditional Medicines, such as Ayurveda–eminently succeed in discovering and applying holistic, and thus effective and safe, treatments by extending their inquiry to the subtle level of Man and Drug by the methods of Intuitive and Analogical Philosophy, while Western Medicine utterly fails to find holistic drugs despite accepting the importance of holism.

True Cosmologies, in the sense indicated above, beget sound sciences of Nature. The Geography alluded to by Ibn Sina (Avicenna) is matter-of-factly global in a very satisfying and fruitful manner, as illustrated by the example of Traditional Medicine. The existing sciences, uncomplemented and uncorrected by Philosophy, still possess great value. However, they have to be complemented by Philosophy, Religious Ethics, Cosmology, and Traditional Sciences etc. Further, they have to be critically judged by Metaphysics and Philosophy, not only for supplying ethical value, but also for correctly distinguishing scientific fact from hypothesis and for determining the acceptability of these hypotheses.

ECOLOGY IN CHANGING DEVELOPMENT PARADIGMS

P. C. Aniyankunju

All production sectors have begun to internalize a new awareness that ecology, human life and development are very much related. This awareness has grown to that level that no development effort can simply overlook the ecological impact of a developmental initiative. But this attitudinal and organizational change has never been instantaneous.

Over the ages, the concept of life and development has changed: from adaptation and/or 'minimal disturbance' to nature in the hunter-gatherer stage to one of ecological "exuberance". The duel with nature, which marked the technological-industrial phase of the development saga, gave rise to ecological disturbances that posed a grave predicament for future generations.

In food production, the above paradigm resulted in the Green Revolution (GR) strategy. Though it has been fairly assumed that GR was a major breakthrough in agricultural production, it does not justify the damage the strategy has done to the future production system. Factors like increased dependence on chemical inputs, land degradation, pest resurgence, destruction of communities of natural enemies, disturbing increase in human poisoning-all lowered economic returns and increased social problems for farm communities.

Development concerns overlooked ecological concerns. A proliferation in the literature on the environmental consequences of modern economic growth and development [Rao, 1974; Kannan and Pushpangadan, 1988; Shiva, 1991; Sharma, 2001] which recognized the delicate relationship between environment and economic relationship gave a foreboding that ruthless pursuit of growth could cause untold irreversible damage to the environment and threaten sustainability of development.

As part of the new awareness, the last few decades have witnessed new social movements and technological initiatives in agriculture to tide over this crisis. The changes that occur in the agrarian scene, as already referred, are due to a change in the perception of life, environment and development. It is argued that a revamping of the lost significance of human life has been taking place in the development thinking [Brinkman, 1981]. Man regains the status as the goal of the development process, which he lost on the way of modern development, than as a tool or factor.

This changing scenario has been a global phenomenon and not confined to Kuttanad[1] the specific region of our study, Kerala or India. In order to understand the changing trends, we need to explore the process of development of agricultural technology. In that process, there has been a drastic change in the attitude towards nature. The concept and the respect for ecology varied over the phases. Correspondingly, we argue that, there has been a paradigm shift in the organization and technology of production.

We propose to examine the changing paradigms in the traditional, modern or Green Revolution and the Post-Green revolution phases in agriculture. We identified the traditional extending from the very spread of agriculture up to the middle of the 1960s: again divided into early phase and pre-modern; the Green Revolution (1968-1990); and the post-Green Revolution, beginning with the early 1990s. However, this is not a watertight division. There is temporal and geographical overlapping. We argued that the respect for nature and therefore the concept and pattern of development of agricultural technology varied across the phases.

Regarding methodology, we interviewed a number of senior and progressive farmers and farm workers from Kuttanad, scientists and agricultural officers during the cropping seasons of 2001-02 and 2002-03. Detailed schedules were used. A primary study was conducted during the farming season of 2002-03, covering all the farmers in the GALASA project.[2]

Theoretical Approaches

Technology per se has been perceived and conceptualized in economic literature from a mere tool to a 'gestalt'[3] and therefore, the relationship between technologies; society and development also vary accordingly. The neo-classical production function approach of technology definitely highlights what the engineering aspect of technology can achieve in modern production, but it disregards the environs in which technology operates. Though technology forms the core of the mode of production, which defines the basic character of social formation, the Marxian approach[4] like the production function approach disregards the impact of growth efforts on environment and sustainability. The Gandhian approach [Narayan, 1968] is deeply rooted in the local environment, but seldom appreciates the potential of modern technologies. However, the Systemic approach [Meier and Baldwin, 1957, Brinkman, 1981] locates technology in its cultural moorings- technology not only in its engineering sense but encompasses the general ethos of the society. Now we proceed to an outline sketch of the development paradigms in agriculture, represented by the technological regimes in their historical setting.

Traditional Agriculture

Traditional agriculture, understood as an indigenous system that employs a crude technology for farming operations is a very lengthy stage, which we define, extended up to the advent of the Green Revolution in the late 1960s. It is divided into two phases; the early and the pre-modern. The early phase refers to the period from the spread of agriculture to the beginning of the second half of the 19th century. The pre-modern phase begins with the second half of the 19th century during which period various reform measures[5] were initiated in the State, which brought about far reaching changes in the socio-economic conditions of the people.

In a way, mechanical technology was more prominent in the early phase of traditional agriculture. Tools and implements, which are central to mechanical technology, were developed and improved in agreement with the topographical characteristics of the area.

In the traditional phase, nature was treated as mother and man, her caretaker. The 'art of agriculture' [Marshall 1890] took shape with the interaction of man with nature. Religious, environmental, social and cultural values shaped the organization and technology of production. Man was the centre of all social and economic activities. Production process was more personal. The wisdom he acquired through his life and experience was inherited by the future generations. Agricultural practices in a traditional society were based on the knowledge of the farming community. Codification of the traditional agricultural knowledge, if any, contained the purity, integrity and prosperity of genuine local knowledge. *Krishigeetha,*[6] a typical example of the indigenous knowledge of Kerala primitive agriculture presents a wholesome knowledge on farming.

Farmers fabricated a variety of tools or implements. The wooden plough tipped with iron, the simple harrows like *palli* and *palaka* (leveling boards), different types of spades, baskets made of bamboo and screw-pine and the ancient sickle were some of the important indigenous tools employed in rice cultivation in Kuttanad. These tools varied across places and farms. Technological diversity is an identifying feature of tradition agriculture. It is explained by differences in local environment or farm variability.

The traditional tools were made of local materials and expertise. Their shape and form were determined by the requirements of the locality and the time. Tool making was totally a local and an endogenous process. Primitive technological activity was linked with the acquisition of means of subsistence and the tools evolved during that process. Subsistence was the driving force (priority) behind the process of technology development.

Traditional farming systems were characterized by locally developed knowledge called indigenous knowledge (IK). Traditional farmers inherited a rich collection of IK, which were handed down through generations by word and practice. *Krishigeetha,* is a codification of the indigenous knowledge of Kerala primitive agriculture, which presents a wholesome knowledge on farming. Also the concept of *njattuvela*[7] is a clear and good example of the scientific approach of traditional farmers. The respect for seasons was central to traditional agriculture and seasons formed an important element of environment.

Cultivators developed farming practices duly respecting nature, without disturbing the ecological balance of the area and respecting the moral right of the future generations to survive. They never tried to correct the environment to suit a particular technology; instead, they tried to develop a technology to suit the environment. The basic culture of mankind had grown hand in hand with agriculture. Religion also exerted considerable influence on the productive system of traditional agriculture. Certain major developments in the pre-modern phase led to the emergence of capitalist farming in Kuttanad that initiated changes in the agrarian relations. The formation of peasant movements and trade unions contributed in their own way in the growth towards modern capitalist farming.

Green Revolution

Green Revolution marks a significant phase in the history of development of agricultural technology. As a strategy Green Revolution meant a package for increasing agricultural production and productivity by overcoming the environmental, cultural and religious constraints on technology. It has been the result of scientific research and systematic application of improved agricultural knowledge especially in the biological and chemical aspects of agricultural technology. The perception of technology was such that it came to be treated as a perfect substitute for nature. It relied upon scientific research and the systematic application of improved agricultural knowledge especially in the biological and chemical aspects of agricultural technology. It ignored the indigenous knowledge and followed a 'blanket technology'.

Green Revolution in Kuttanad happened to be a revolution in the biological and chemical aspects of agricultural technology. Miracle seeds or HYVs have been central to the strategy. The second important practice typical of the GR technology is the increased application of inorganic fertilizers and plant protection chemicals (PPCs).

One of the main issues related to the ecology of Kuttanad has been the excessive use of chemical fertilizers and pesticides [George and Krishnakumari, 1993], which increased the occurrence of more pests and diseases. HYVs low tolerances to pests and diseases and non-judicious land protection measures have increased the incidence of pests [John et al, 1990; Thomas et al, 1990; George and Krishnakumari, 1993]. Crops, which were effectively protected with four to five treatments per season, had to be sprayed ten to fifteen times in subsequent years [Narayanan, 2003]. The dosage had to be increased every year. A farmer shared his experience that there was a steady decline in milk production of cows and related to it the pesticide pollution of the water in the canals. An inland fisherman recollected the heavy incidence of disease on fish population in the 1980s in Kuttanad.[8] It also affected the fish catch from the inland waters, which declined sharply, and the popular species like prawns disappeared. Pesticide use has seriously affected the eco-milieu of Kuttanad, especially after the wide spread destruction of the frog population [Narayanan, 2003].

The consequences of Green Revolution have been far reaching. The central element to the strategy has been a particular perception of technology, in which technology was considered as a perfect substitute for nature. The consequences of the GR were such that it caused a worldwide awareness against the unsustainable aspects of the New Strategy. New initiatives to balance the goals of cost minimization, sustainability and increased production began to be perceived and attempted. This marked the post-Green Revolution phase.

The Post Green Revolution:

The post Green Revolution phase has been characterized by a wide range of alternative farming practices or approaches to the bas agrarian question of organizing production, emerged as a response to the consequences of the GR strategy. Most of these alternatives were in the form of sporadic and scattered initiatives.

The term sustainable agriculture synthesizes a variety of concepts and perspectives associated with agricultural practices. It is management system for renewable resources including soil, wild life, forests, crops, fish, livestock, plant genetic resources and eco system to provide food, income and livelihood for current and future generations and that maintains or improves the economic productivity and eco-system services of these resources.

While the traditional agriculture served the goals of subsistence and harmony with nature, the Green Revolution agriculture aimed at maximizing productivity and production, unmindful of the social and environmental consequences. In contrast, sustainable agriculture requires the balancing of a variety of goals. Therefore, one of the basic concepts of sustainable development and sustainable agriculture is the concept of trade-offs. Trade-offs imply policy and individual decisions regarding such factors as current versus future income, private versus public benefits and future growth versus current consumption. This approach of trade-offs would necessarily influence the pattern of technology development.

In Kuttanad, as elsewhere, there has been a growing awareness among administrators, scientists, extension personnel, farmers, social workers, NGOs and farm labourers that the system is unsustainable and thereby the farming practices need to be reexamined and alternatives evolved to avoid further decline in soil health and environment. They include individual initiatives in organic farming, integrated farming system of 'one rice-one fish' farming. Another initiative has been the GALASA. It is synthesis of three elements of a successful agricultural development strategy. Group Approach (GA), Locally Adapted (LA), and Sustainable Agriculture (SA). It has been a combined effort based on the principle of communion among farmers, farm workers, extension workers and agricultural scientists. It involved the use of organic manure, proper land preparation, use of HYV of seed, minimum dependence on chemical inputs and changed plant protection measures. Thus, over the three phases, there has been drastic changes in the technology of agricultural production.

We discuss the agricultural development experience in the Kerala case and identify the pattern of the process of development of agricultural technology. We find that, over the three phases of agricultural development, the agents and priorities of technology development have changed. The particular perception of ecology, life and development of the agents and their priorities defined the nature of technology developed and its impact.

Pattern of Technology Development in the Traditional Phase

In the traditional phase, the farming community or society played an important role in the production of indigenous knowledge respecting the local environmental, religious, social, and cultural values. Each farmer depending on his ingenuity and creativity added to or improved the stock of knowledge. The farmer might be an owner cultivator, tenant cultivator or even a farm labourer. It was in his capacity as the actual tiller that he contributed to the process of technology development. The farmer as the agent of technology development drew much from the society. He inherited a strong sense of values in and through the society. In his efforts as the agent of technology development he was guided, tuned or limited by the above values.

Development of technology in traditional agriculture followed a particular pattern. It was a cultivator/society-driven one. It followed a path of trial and error. Countless efforts of primitive producers to advance agricultural technology contributed to the stock of farm know-how. These efforts to enlarge the reproduction process under the growing scarcity of natural resources seem to have been induced by the producers' need for sustenance [Hayami, 1997:12]. The priority of the farmer was his sustenance and of his family. Whatever surplus left over after subsistence, he exchanged it with the members in the local community.

Farmers had been dedicated plant and animal breeders for thousands of years, although not in the precise manner of modern genetics. They have consciously maintained diversity, planted mixed fields systematically to achieve natural crosses, practiced selection and set up their own personal gene banks as well as far-flung exchange systems for acquiring new genetic material.

In all the farming operations the farmers inherited a stock of knowledge from their forefathers, and depending on their creativity and ingenuity, added to that stock, giving a flow character to the knowledge. In his traditional role as the chief agent/actor of agricultural technology development, the traditional or the farming community made significant contribution to the farming system.

Regarding the priorities in technology development in traditional agriculture, it was the farmers who set them. Traditional agriculture was dominated by an attitude, which emphasized survival and maintenance of position rather than improvement and advancement of position. They followed farming for sustenance. They sought such a technology to suit the environment.

Exploitation of nature and commercialization of production were not the priorities of traditional farming. The farmers wanted to be friendly with the nature and did not want to disturb the harmony of men with nature. They developed the technology in such a way as to suit the particular environment, in contrary to the pattern of developing a system to make the environment to suit a particular technology. For instance, the farming system that they developed respected the rhythm in the nature. The centre of *njattuvela* was that rhythm. They planned the timing of the operations and the type of crops to be raised according to the set annual pattern of variations in the nature. They set the priorities respecting the limits predetermined by the environment.

The above discussion on the pattern of technology development in traditional agriculture is summarised in Exhibit 1.

Exhibit 1: Location-specific Cultivator/Society-driven Technology Development System in Traditional Agriculture.

The basic orientation of this pattern of technology development is that there exists sincere interactive relationship between individuals and the society, shown by the directional arrows. Each individual as a cultivator and as a traditional innovator receives various environmental, cultural, religious and social values in and through the society. His entire life style is based on these values and he respects them.

When he looks out for new ways of doing things his basic orientation to these values does not change. Though his innovative efforts are marginal, he is one of the links in the chain of technology development and its intra and inter-generation of transfer. He contributes to the stock of knowledge of the farming community, shown by the arrows from the cultivator to the society. The technology developed and transferred in a society embedded in its environmental, cultural, religious and social values which makes the process purely an endogenous one.

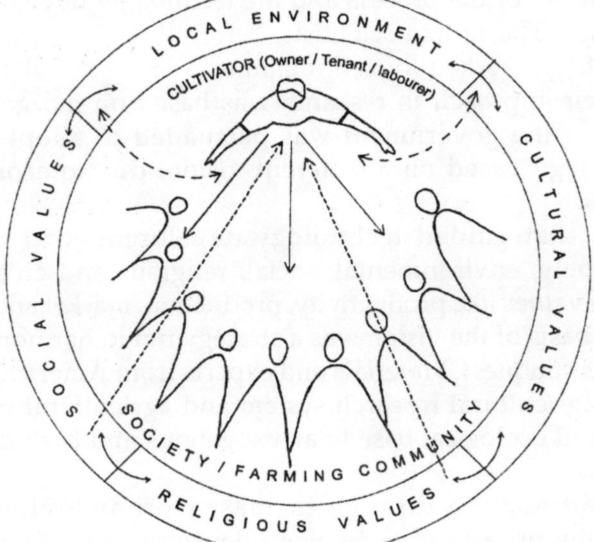

To sum up, in the development of technology in traditional farming, the cultivator acted as the chief agent/player. Farm and the homestead were the fields of his experiments. However, he was controlled and/or tuned by the particular social, religious, cultural factors and the environment in which he lived. The Green Revolution strategy of technology development, however, makes a radical change in the agents and priorities and has its consequences.

The Pattern of Technology Development in the Green Revolution Phase

The orientation and the organizational pattern and the process of technology development in the Green Revolution phase were altogether different from those prevailed in the traditional system. There took place a paradigm shift in the agents/actors, forces and the priorities of technology development.

The Green Revolution technology, as explained, was developed against the background of severe food scarcities and famines. The third-world countries imported and adopted the 'package' for addressing the problem of food deficit. But the process of adoption did not change the basic orientation of the imported technology. Everything was orientated towards increasing output. Projects aimed at immediate expansion of production took high priority. Thus a technology, which was highly responsive to inputs, stood at the centre of the Green Revolution strategy. Hence, the 'miracle seeds' or the High Yielding Varieties became the hallmark of the new technology. Research has been central to the pattern of technology development in this phase.

Technology development or research at any time has been value oriented. The nature of the process and the technology developed varied with these values. The Indian scientists, before the advent of the Green Revolution strategy, worked for self reliant and ecological alternatives in production. Their approach in research was based on indigenous roots [Jha, 2001:2]. But the government was persuaded to adopt the Green Revolution strategy based on a different vision, due to economic and political reasons.

The values that guided technology development in traditional agriculture, namely, environmental, social, religious and cultural, were replaced by new values like productivity, production, market and economic prosperity. The basis of the vision was a strategy not in harmony with the nature, but on its conquest. Materials and experts from America were used to shift India's agricultural research system and agricultural policy from an indigenous and ecological base to an exogenous and high input-using system.

Increased production through increased productivity has been highlighted as the priority of technology development. The orientation has been to develop a technology to intensify the exploitation of the agricultural base in developing countries. Such a technology warranted increased amounts of non-renewable resources drawn from outside the agricultural sector, e.g. inorganic fertilizer.

The scientists, supported by the political decision and funding offered a technological solution to scarcity and poverty. They developed a technology to enable the farming community to tide over the food crisis. The scientist assumed the chief role as the agent/actor of technology development. His sole priority has been to increase production and productivity.

Thus technology has become an exogenous variable, an outsider. Technology to be followed in a locality came to be developed outside the system. The traditional role of the farmer and his community in the origin and development of a system of production came to be totally ignored. Researchers and technologists in institutionalized organizations developed the technology that they perceived to be important and eventually transferred them to farmers through extension personnel. This approach totally disregarded the traditional role of farmers in moulding agricultural technology and bypassed the traditional knowledge base of the farming communities. The farmer has become a technology taker than a maker. The location specificity of technology application suffered in the process. It has been assumed that the success of the farmer depended solely on the extent and the sincerity he maintained in adoption. For poor performance, if any, the farmer came to be blamed for his low rate of adoption. Exhibit 2: Green Revolution, Scientist-driven Technology Development and Transfer System.

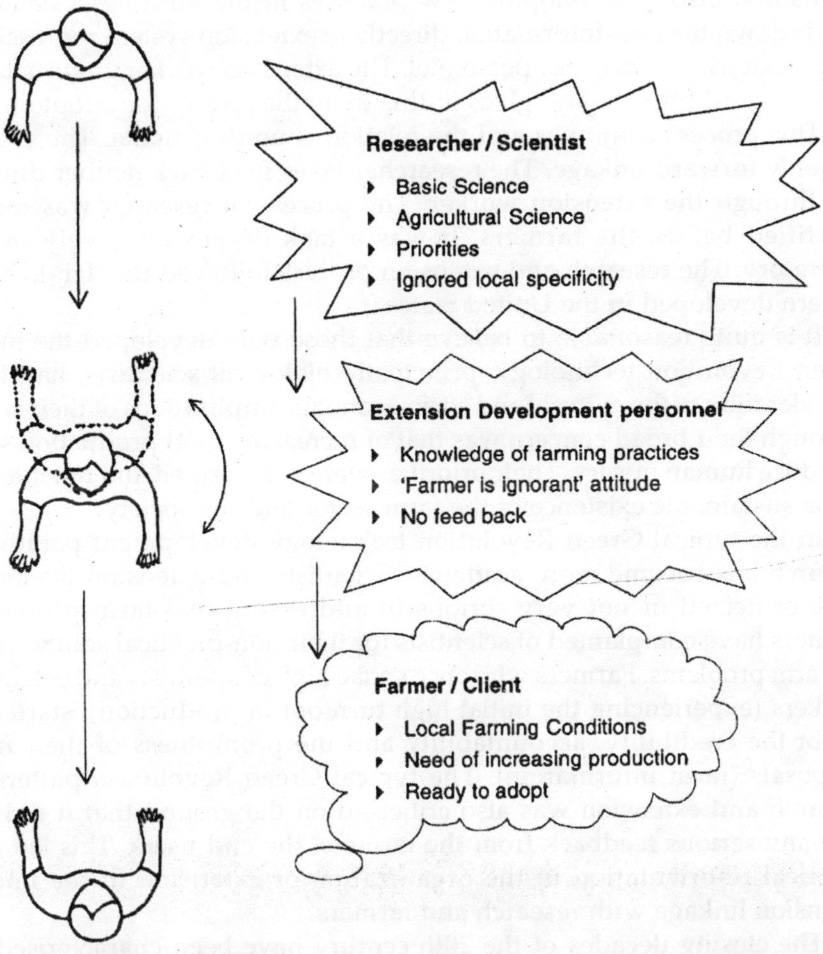

Peasants as plant breeding specialists gave way to scientists of multinational seed companies and international research institutions like CIMMYT and IRRI, national and regional research stations.

The upper hand the scientists enjoyed in seed breeding extended to the farming systems. The shift from indigenous varieties of seeds to the Green Revolution varieties involved a shift to a farming system controlled by agro-chemical and seed corporations, and international agricultural research centres. In the Green Revolution technology development process scientists took the full effort and credit as they performed as the chief agents or actors in the process. The following exhibit summarises the pattern of technology development in the Green Revolution system.

The scientist or researcher represents the whole system of technology development. The impression has been that he has sufficient motivation in the process and develops the technology that he thinks fitting to the farmer. He conducts research on new varieties of seed and other farm inputs and also in developing new practices in the farming system. He hands down the new information directly to extension system represented by the extension worker or personnel. The extension worker, in turn, takes it for onward transfer and gives it directly to the farmer for adoption.

This process continues and the relation is unidirectional. The system has only forward linkage. The researcher takes feed back neither directly nor through the extension worker. The process of research was totally mystified before the farmers. It was a task done exclusively in the laboratory. The research and extension process followed the 'lab-to land' pattern developed in the United States.

It is quite reasonable to believe that those who developed the initial Green Revolution technology, principally biological scientists, had little consideration to the cultural and socio-economic implications of their work, although their broad concern was that of increasing food production so as to reduce human misery. Their priorities were very limited and not integral to the sustainable existence of the farm sector and the society.

In the typical Green Revolution technology development paradigm, research has become more academic. Scientists and extension personnel were criticized of not very serious in addressing real farm problems. Farmers have complained of scientists for their non-practical solutions for the farm problems. Farmers who once spoke high of scientists and extension workers (experiencing the initial high turn out in production), started to doubt the credibility, accountability and the promptness of their farm proposals (farm information). The typical Green Revolution pattern of research and extension was also criticized on the ground that it did not take any serious feedback from the farmers, the end users. This led to a practical re-orientation in the organization of extension in the line of extension linkage with research and farmers.

The closing decades of the 20th century have been characterised by serious rethinking on the continued ability of the Green Revolution strategy to feed the future population. Many warned against the dependence on the Green Revolution technology that endangered the very existence of humanity. Therefore, serious efforts have been coming up, in not only reducing the consequences of the Green Revolution, but in evolving alternative technologies. A significant trait of this new trend has been the increasing tendency of accepting the creative role of farmers in technology development and diffusion. He has been made a participant in the process. There has been an increasing global call for the same.

Post-Green Revolution agriculture is recognizing the need to incorporate previously underutilized farmer experimentation [Box, 1987; Rhoades, 1989], farmer participation in formal research [Ashby, 1987; Biggs, 1988; Farrington and Martin, 1987] and in technology development [Malton et al, eds. 1984; Sagar and Farrington, 1988]. The pattern of technology development concepts like appropriate technology, technology blending, participatory technology etc., have evolved. New environment-friendly systems and practices like Farming Systems Research, On-farm Research, Integrated Pest Management (IPM), Integrated Nutrient Management (INM), Integrated Client-Oriented Research Projects etc., have been developed and popularised. Correspondingly, new initiatives like Organic Farming, Integrated Farming Systems (IFS) like, 'one rice-one fish' cropping, GALASA etc., are being tried. All these could be termed as Post-Green Revolution initiatives, as they marked a deviation from the Green Revolution pattern and process of technology development.

In Kuttanad, the Operational Research Project (ORP)[9] has been a response to the environmental consequences due to increased application of chemical inputs. All theses programmes were with the participatory role of the farmers. In the pest and disease surveillance programme farmers were given timely warning on the preventive and control measures to be taken based on the counts of insect pests collected in the light traps, and observation from the fixed plot and roving survey.

This feedback function of extension facilitated the continuous reorientation of research towards the priority needs of farmers and the early resolution of important technological constraints [Benor et, al 1984: 7]. The implication of T & V system to technology development has been that the farmers came to be heard by the scientists directly or through the extension feedback in the process. The research system began to address the practical problems raised by the farmers regarding adoption. The system tried to ensure link between scientist/research and farmers, as shown in the following exhibit.

This created an atmosphere of participatory technology development, in which the scientist or researcher based on his knowledge of basic and agricultural science received feedback from the extension workers and the farmers. He is influenced by the feedback and as the chief agent in technology development sets the priorities and is aware of the challenges in the field and research.

As per this pattern, the extension or development worker is aware of the local farming conditions and he has the information about the farming practices, received through training. There has been a significant attitudinal change on the part of the extension worker towards the farmer. He came to treat the farmer as sensible. Certain farm practices that have been developed and followed by farmers came to be accepted and recommended by the scientists and the extension workers. A typical case is the practice of re-flooding for weed control before sowing [Thomas, 1994].

Exhibit 3: Transitional, Participatory Technology Development and Transfer System

The farmer on the other hand, began to keep a critical attitude towards the information passed on by the extension worker regarding farming practices. He was not ready to fully adopt the recommendations of the extension worker. He gave his feedback to the extension worker or to the researcher and thus influenced the research process. A degree of participation was ensured in the pattern.

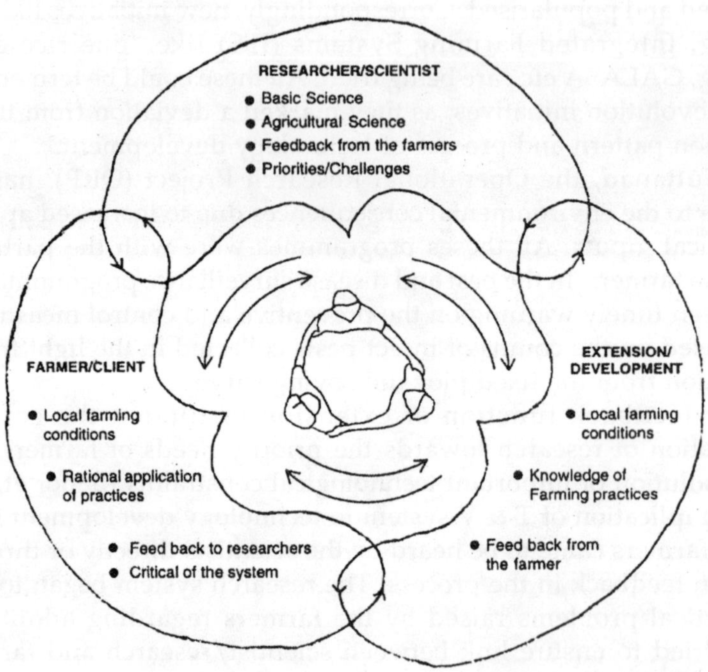

Thus, a new system of technology development is evolving in Kuttanad. A re-orientation among the agricultural scientists and extension personnel has been taking place to develop a productive and inter active relationship among the various agents in technology development. Time tested traditional farm practices developed by the farming community, as mentioned before, are appreciated and popularised by researchers and extension staff. Scientists and extension staff are becoming more receptive to farmers. The local governments through funding promote all genuine, sustainable farmer initiatives in technology development. In the whole process NGOs come to the help of the farming community in motivating the farmers.

In each phase of agricultural technology development, there were elements of the previous phase, but we identified each phase in terms of the new pattern of technology development that had emerged in each phase. Therefore, it has been observed that in each phase, at any particular moment of time, we come across elements of the previous phase resulting in dualism [Raj, 1990].

To sum up, the pattern of technology development has changed across the three phases. The decisive factors in the process seem to be the particular socio-economic perception of the agents and their priorities. The consequences of a technology depended on the above factors. As has been argued earlier, remnants of the previous system are seen in every phase. However, the phases are marked by significant changes in the organization of production and technology.

Thus, technology has been integral to the socio-economic system; it defined the system. Several factors- environment culture, socio-economic structure and concern for the future- significantly influenced the nature of technology developed with varying importance in the long history of technology and economic development. Each technology regime as embedded in a unique socio-economic system-ethos. Hence in each of the socio-economic technology regime, agents and priorities in technology development also changed in accordance with the approach towards nature and the dominant ethos of the system.

Notes

1. Kuttanad was the principal rice-growing region in the former princely state of Travancore. It has been a unique heritage of humanity with natural beauty and enterprising people. Paddy cultivation in Kuttanad has been a test of great human effort. The special features of Kuttanad make the method of paddy cultivation unique. It is one of the handful areas in the world following wetland agriculture below the sea level, by 0.6 to 2.2 meters.
2. Group Approach (GA), Locally Adapted (LA) and Sustainable Agriculture (SA), a post green revolution initiative in agriculture initiated in the district of Palakkad and in Kuttanad.
3. Gestalt is a word taken from psychology and is used to mean figure formation, a holistic configuration that determines all its parts.
4. Marx summarized the relation between technology, man and nature thus: Technology discloses man's mode of dealing with Nature, the process of production by which he sustains his life, and thereby also lays bare the mode of formation of his social relations, and of the mental conceptions that flow from them.[Karl Marx , *Capital,* Vol. I, Modern Library edition, p. 406.]
5. Reforms such as the Pattom Proclamation, abolition of slavery; developments like introduction of money wages, demand for marketable surplus; mechanization of certain farming operations, formation of trade unions and governmental interventions like research, extension, formation of Industrial Relations Committee (IRC), construction of engineering structures like Thottappally Spillway and the Thannermukkam Regulator and Kayal Reclamations.
6. *Krishigeetha* is a codification of the traditional agricultural knowledge. There are different versions. Generally it speaks of seed and its variety, geography, climatic conditions, crop diversity, weeding methods, manure and its application, agricultural tools, drought animals, timing of various farm operations, moral advises and moreover a life style based on farming. It is typical example of the indigenous knowledge. It presents a wholesome wisdom on farming.

7. *Njattuvela* speaks of the best time for each farming operation especially transplanting seedlings. This idea was at the core of the primitive agricultural strategy. The respect for seasons was central to traditional agriculture. Seasons formed an important element of environment.
8. Information collected from the field through interaction with the farmers and inland fishermen on 02-05-2003.
9. It was a joint venture of the Kerala Agricultural University and the Department of Agriculture. It included projects like Integrated Pest Management (IPM) with pest and disease surveillance and adaptive trails and Integrated Nutrient Management (INM) with fertilizer trials.

References

Ashby, J.A. (1987), 'The Effects of Different Types of Farmer Participation on the Management of on Farm Trials'. *Agricultural Administration and Extension* 24: pp. 234-252.

Benor, Daniel, James Q. Harison and Michael Baxter (1984), Agricultural Extension The Training and Visit System A World Bank Publication, Washington D.C, USA.

Biggs, S.D. (1988), Resource- Poor Farmer Participation in Research a Synthesis From Nine National Agricultural Research Programs. The Hague: International Service for National Agricultural Research.

Box, L. (1987), 'Experimenting Cultivators: A Methodology for Adaptive Agricultural Research, London: Overseas Development Institute (ODI) Agricultural Administration (Research and Extension) Network Discussion Paper 23.

Brinkman, L. Richard (1981), *Cultural Economics* Heritage Publishers, New Delhi.

Farringson, J. and A. Marhin (1987), 'Farmer Participatory Research: A View of Concepts and Practices' Agricultural Administration (Research and extension) Network Discussion Paper.19. London .ODI.

George, Alexander and P.R. Krishnakumary Amma (1993), 'Trends in the use of Ecologically Hazardous inputs for rice in Wetland ecosystem: A 10 year case study in Nair R.R., K.P. Vasndevan, C.A. Joseph (eds.) *Rice in Wetland Ecosystem*, Kerala Agricultural University, pp. 250 to 253.

Hayami, Yujiro (1997), *Development Economics From the Poverty to the Wealth of Nations*, Clarendon Press, Oxford.

Jha, Dayanatha (2001), 'Agricultural Research and Small Farms' *Indian Journal of Agricultural Economics*, Vol. 56, No.1, January to March, pp. 1 to 24.

John, *et al.* (1990), 'Effect of Agronomic Practices on Rice Yield and the Incidents of Major Rice Pests' in *Rice in Wetland Ecosystem*. Kerala Agricultural University in Thrissur.

K. Kannan, K.P. and Pushpangadan, K. (1988), 'Agricultural Stagnation in Kerala: An Exploratory Analysis' in Economic and Political Weekly, Review of Agriculture, Vol. XXXIII, No. 39, Sept.

Marshall, A. (1890), *Principles of Economics* Eighth Edition. London 1969.

Narayanan, N.C. (2003), Against the Grain The Political Ecology of Land use in a Kerala Region, India Published Ph.D.

Raj, K.N. (1990), *Organizational Issues in Indian Agriculture*, Oxford University Press, New Delhi.

Rao, VRRV (1974), *Growth With Justice in Asian Agriculture*. An Exercise in Policy Formation, United Nations Research Institute for Social Development. Geneva.

Robert, Rhoades (1989), The Role of Farmers in the creation of Agricultural Technology in Chambers Robert, Arnold Parcy and Lori Ann Thrupp (eds.) *Farmer First Farmer Innovation and Agricultural Research, London*.

Sagar, D. And J. Farrington (1988), Participatory approaches to Technology Generation: from the development of methodology to wider-scale implementation. London: ODI. Agricultural Administration (Research and Extension) Network Paper, 2.

Sharma, Sudhir (2001), 'Green Revolution and Soil Nutrient Depletion in Punjab: is there a reason for Concern?' (Working Paper) Institute of Economic Growth, Delhi University.

Shiva, Vandana (1991), *The Violence of Green Revolution*. Third World Agriculture, The Other India Press, Goa. 1991.

Thomas, T.A. and Thomas J.A. (1999), 'Changing Agrarian Relations and Practice in Kuttanad in M.A. Oommen (ed.) *Kerala's Development Experience: Global and National Dimensions*, Vol. II, Concept, New Delhi.

ECOLOGICAL SPIRITUALITY IN CROSS CULTURAL PERSPECTIVE-ITS LEGAL DIMENSIONS

Shobha Menon

Nature and Religious Ethos

Is conservation of Nature in our religious Ethos? I shall briefly touch upon the subject before examining the legislative measures adopted by various bodies to preserve ecology for the survival of man. The ancient Hindus were nature worshippers. Hindu Deities of Vedic period were symbols of natural forces without which life could not have been sustained. Agni (Fire), Varuna (Water), Vayu (Air) and Surya (Sun) were some of the Chief Deities to whom a number of hymns in praise were dedicated in the Vedic Scriptures. Even during the period of Puranas, though other Gods were added, the deities signifying natural forces remained as important as before. Various Hindu Gods had their own transportation measures based on animals and birds. Lord Siva used the Bull, Durga a Tiger, Kartikeya a Peacock, Ganapati a Mouse and Saraswati a Swan. Well; this concept helped in preserving the animals and birds as these came to be revered by the Hindus.

According to Bible God created Earth and the flora and fauna first and then the man in his own image. And God blessed them and God said unto them, "be fruitful, and multiply and replenish the earth". Conservation was implied here and not wanton destruction though man was put in command of the natural resources that God had created. However 'Ahimsa' is the Corner stone of Buddhism and Jainism. Compassion for the fellow beings was an important facet that forbade the followers from killing other living beings. Ecologic balance was a natural consequence that followed.

Men were Wiser before Modern Civilization Set in

In the year 1854, "The Great White Chief in Washington" made an offer to "The Great Wise Indian Chief of Seattle" to buy their land. The reply sent by the Indian Chief was profound. It contained the wisdom of Ages. It was the first ever and the most understanding statement on environment.

The full text was published in "Paryavaran Vol. I, No: I June 1984". It is worth quoting extracts from the letter.

"How can you buy or sell the sky, the warmth of the land? The idea is strange to us.

If we do not own the freshness of the air and the sparkle of the water, how can you buy them?

Every part of the earth is sacred to my people. Every shining pine needle, every sand shore, every mist in the dark woods, every clearing and humming insect is holy in memory and experience of my people. The sap which courses through the trees carries the memories of the red man."

We are part of the earth and it is part of us. The flowers are our sisters; the horse, the great eagle are our brothers. The rocky crests, the juices in the meadows, the body heat of the pony, and man all belong to the same family.

We know that the white man does not understand our ways. One portion of land is the same to him as the next, for he is stranger who comes in the night and takes from the land whatever he needs. The earth is not his brother but his enemy, and when he has conquered it, he moves on. He leaves his father's grave behind, and he does not care. He treats his mother, the earth and his brother, the sky, as things to be bought, plundered, sold like sheep or bright beads. His appetite will devour the earth and leave behind only a desert. There is no quiet place in the white man's cities. No place to hear the unfurling of leaves in spring, or the rustle of an insect's wings. But perhaps, it is because I am savage and do not understand. The air is precious to the red man, for all things share the same breath-the beast, the tree, the man; they all share the same breath. The white man does not seem to notice the air he breathes.

So we will consider your offer to buy our land. If we decide to accept, I will make one condition. The white man must treat the beasts of this land as his brothers. What is man without the beasts? If the entire beast were gone, man would die from a great loneliness of spirit. For whatever happens to the beasts soon happens to man. All things are connected. This we know: The earth does not belong to man; man belongs to the earth. Whatever befalls the earth befalls the sons of the earth. Man did not weave the web of life: he is merely a strand in it. Whatever he does to the web he does to himself.

Today's society's interaction with nature is so extensive that the environmental question has assumed proportions affecting all humanity. Industrialization, urbanization, explosion of population, over-exploitation of resources, depletion of traditional sources of energy and raw materials and the search for new sources of energy and raw materials, the disruption of natural ecological balances, the destruction of a multitude of animal and plant species for economic reason and sometimes for no good reason at all are factors which have contributed to environmental deterioration.

Let us not, however, flatter ourselves overmuch on account of our human victories over nature. For each such victory nature takes it revenge on us. Each victory, it is true, in the first place brings about the results we expected, but in the second and third place it has quite different unforeseen effects which only too often cancel the first. Ecologists are of the opinion that the most important ecological and social problem is the widespread disappearance all over the world of certain species of living organisms. Biologists forecast the extinction of animals and plant species on a scale that is incomparably greater than their extinction over the course of millions of years. Nature and history, it has been said, are two component parts of the environment in which we live, move and prove ourselves.

The Genesis of Ecological Degradation

At one time there was a perfect ecological balance of nature. With the growth of economic power and unplanned industrial development, the pressure on natural resources has increased considerably. In his lust for comfortable life, man has miserably built up a high degree of pressure on natural resources causing thereby a danger for human well being and health. Perhaps man has forgotten that the depletion of natural resources may lead to the destruction of human that his duty is to add something to the natural wealth in consideration of what he receives from nature. All this has resulted in an eco-catastrophe. Due to man's increased interference with the nature, the problem of environmental pollution has increased manifold. The problem of environmental pollution has assumed international dimension. Ecology is the science of the intricate web of relationships between living organisms and their living and non-living surroundings. It is a branch of biology dealing with relations of living organisms to their surroundings.

All the environmental problems associated with development have raised several questions regarding the type and nature of development and this has given rise to the concept of sustainable development. Sustainable development means a kind of development that meets the needs of the present generation without compromising the ability of future generations to meet their own needs. The various programmes of UNEP, have emphasized the need for sustainable development. The concept of sustainable development has dimension-environment conservation. Sustainable development is one which conserves the nature and maintains ecological order-biodiversity as it is commonly referred to-thereby making life on earth possible in future as at present. But a problem of making human beings live in better harmony with nature than they are currently tending to.

Sustainable Development as a balancing concept between ecology and development has been accepted as a part of customary international law. Some of the salient features of sustainable development are intergenerational equity, use and conservation of natural resources, environmental protection, the precautionary principles, and polluter pays principle, obligation to assist and co-operate eradication of poverty and financial assistance to the developing countries.

Combating Pollution

Imbued with greed and the appetite for material powers we embarked upon adventures in the name of development, detrimental to our interest, Man treaded the path of development and industrialization according to his own perceptions unmindful of its pernicious affects on environment.

The factories, automobiles, solid and liquid waste disposal, fast moving traffic, rites-rituals celebrations by bursting crackers, displaying amplified orchestra, band and loudspeakers and by disposing of dead bodies into the rivers, we are causing land, air, noise and water pollution, heading towards a disaster. The multifarious needs, desires and ambitions of man have a direct impact on nature's environment. The saying of Paul Harrison, seem to be correct that "the rape of the earth has been central to the past progress of Westerns industrialism, the plunder and the waste of irreplaceable resources, the poisoning or destruction of fragile ecosystems, the beliefs that man can mould nature with no need to take account of nature's reaction to such a violation. Disdain for traditional cultured and social system was a fundamental characteristic of all modern European empires, and this attitude spread to the new ruling elites who had adopted western ways.

At the national level in India, not only legislative and administrative measures have been taken for protecting and combating pollution but judiciary has also endeavored to curb the menace of environmental pollution and realized its responsibility as shown in its judicial adventurism. The Government of India in 1972 established a National Committee on Environmental Planning and Co-ordination to advice on recommendations for improving the environment to Central Government. The creation of Department of Environment is the direct result of the recommendations of Tiwari Committee appointed by Central Government in 1980 for recommending legislative measures and administrative machinery for ensuring environmental protection.

The need for proper balance between the development planning and environmental management has been acknowledged. In the eighth plan an outlay of Rs. 525/- crores was provided for forestry and wild life sub-sector and Rs. 675/- crores for ecology and environment sub-sector. Apart from administrative measures some legislative enactments are also made to deal effectively with different kinds of environmental pollution. Today we are having more than 200 central and state laws to protect environment. Our Constitution, the supreme law of the land provides for some environmental provisions in the context of our national dimensions of human rights, right to life, liberty and pollution free environment. Article 21 of the Constitution provides "No person shall be deprived of his life and personal liberty except according to procedure established by law".

Article 48-A provides for the protection of environment and safeguarding of forests and wild life. The State shall endeavor to protect and improve the environment and to safeguard the forests and wild life of the country to protect and improve the natural environment including forests, lakes, rivers and wild life and to have compassion for living creatures. The forms of nuisance are innumerable, noise, smells, pollution of air and water are the most usual instances. Excessive tolling of church bells that interfere with personal discomfort, are also the instances of nuisance. We in India have no paucity of anti-pollution laws which require some amendments, but the need of the hour is to implement and enforce these laws rigidly when the legislature is lagging behind in plugging the holes in the existing legal system and Administration is not well equipped to meet the challenge.

The right to life under Article 21 of the Constitution includes right to enjoy pollution free water and air and healthy environment. Ensuring green belts and open spaces for maintaining ecological balance forbidding stone crushing activities near residential complexes earmarking a part of the reserved forest for Adhivasis to ensure their habitat and means for livelihood, compelling the municipal authorities of the Delhi Municipal Corporation to perform their statutory obligations for protecting the health of the community, compelling the industrial units to set up effluent treatment plans, directing installation of air pollution controlling devices for preventing air pollution, directing closure of recalcitrant factories in order to save the community from the hazards of environmental pollution are some of the significant cases displaying judicial activism. Hygienic environment is an integral facet of right to healthy life; therefore, environmental protection is a matter of grave concern for human existence.

The Three Principles of Anti-Pollution

(a) Prevention i.e. new products must be developed in such a way as to avoid pollution or damage e.g. chemicals, besides food, air and water, have always been part of man's environment some measure. The ecosystem has become imbalanced by uncontrolled use, abuse and misuse of natural resources and manufacture and use of hazardous products and chemicals resulting in endangering the very existence of human race. Apex Court has recently held that "though sufficient steps have been taken to either ban or to allow restrictive use of these insecticides yet there is no coordinated effort by different ministries".

(b) The polluter pays i.e. it is not the public at large but those causing the damage or pollution who bear the responsibilities and the cost of removal e.g. The polluter pays principle has been held to be a sound principle by Supreme Court in Indian Council for Environment-Legal Action Vs Union of India, AIR 1996 SC 1446. The Court observed...We are of the opinion that any principle evolved in this behalf should be simple, practical and suited to the conditions obtaining in this country. The court further ruled that.......Once the activity, carried on is hazardous or inherently dangerous, the person carrying on 'such activity is liable to make good the loss caused to any other person.

(c) Co-operation, i.e. government, industry and society join forces to combat environmental problems since every individual has a fundamental duty under Article 51-A (g) of the Constitution towards environment. The Supreme Court in Vellore Citizen's Welfare Forum Vs Union of India, AIR 1996 SC 2715 explained the 'Precautionary Principle' and upheld the Polluter Pays Principle, as "some of the salient principles of "Sustainable Development" as called out from Brundtland Report and other international documents, are Inter-Generational Equity.

International Invitations

(a) Stockholm Conference on Human Environment: The global interest and concern about the environmental pollution has arisen because of the First Conference of Human Environment convened by 24th General Assembly of United Nations Organization at Stockholm in June, 1972.

(b) I.U.C.N. Resolution: I.U.C.N. Resolution was passed in 15th session of the International Union for Conservation of nature and Natural Resources (I.U.C.N.) held in New Zealand in October 1981. The resolution related to environmental conservation and world conservation strategy.

(c) Nairobi Declaration 1982: To celebrate the tenth anniversary of the 1972 Stockholm Conference on the Human Environment, 105 nations gathered at Nairobi from May 10-18, 1982, and adopted the Nairobi Declaration.

At the Nairobi meeting, a decision was taken for the creation of a special commission to propose long term environmental strategies for achieving "sustainable development to the year 2000 and beyond", which was endorsed by the Government Council of the UNEP in 1987 and later by General Assembly.

(d) Rio Summit: In Sept. 1989, the Soviet Foreign Minister proposed for a summit meeting on Environment, in order to adopt some kind of code on environmental behaviors to give a strong push to environmental related activities. Accordingly the RIO summit was held in RIO De Janeiro from June 3, 1992 to June 14, 1992. This summit is popularly known as Earth Summit. The RIO declaration proclaims that "Human beings are at the Centre of concern for sustainable development. They are entitled to a healthy and productive life in harmony with nature. The Earth Summit ended on June 14, 1992 after adopting the Rio declaration. Agenda 21 on the problems of green house effect Agenda 21 contains a checklist of do's and don'ts the rights and obligations to protect.

(e) Berlin Summit: The UN climate conference in Berlin adopted mandate on strengthening the 1992 RIO climate treaty. It obliges developed countries to set objectives for further emission cuts by 1997.

(f) United Nations Commission on Sustainable Development: Following the Earth Summit, the UNCSD in pursuance of the decision of the United Nations General Assembly, established the United Nations Commission on Sustainable Development (UNCSD) to fulfill a range of functions. These functions broadly fell into the following three categories.

1. Monitoring the implementation of Agenda 21.
2. Reviewing the availability of financial and technical resources.
3. Acting as a forum for discussion consensus building and decision making, which will include identifying weaknesses in the international legal and institutional regime.

Presering Environment: Legislative Measures taken in India

Besides Constitutional remedies, there are many other enactments for preserving and protecting the environment in its existing form and control the pollution thereof. Some of the important legislators are mentioned as following:

(a) Fair Way Act, 1881. The Police Act, 1861. Northern Canal and Drainage Act, 1873. Indian Penal Code 1860. Oriental Gas Company Act, 1857. Poison Act, 1919. Indian Boilers Act 1923. Indian Forest Act, 1927. The Air Craft Act, 1934. Factories Act, 1948. The insecticides Act, 19689. The Wild Life Protection Act 1972. Code of Criminal Procedure, 1973. The Water (Prevention and Control of Pollution) Act, 1974 and rules there under. The Water Pollution Cess Act, 1977. The Forest Conservation Act, 1980. The Air (Prevention and Control of Pollution) Act, 1981. Environmental Protection Act, 1986, Motor Vehicles Act 1989.

(b) The environmental pollution causing nuisance is redressable for an action under S. 133 Cr. P.C. The nuisance coming within the purview of civil wrong is redressable by an action through an individual or by an action in the representative capacity under Section 91, C.P.C.

Concern over the state of Environment has grown world wide since the sixties. The decline in environmental quality has been evidenced by increasing pollution. Loss of vegetal cover and biological diversity, excessive concentrations of harmful chemicals in the ambient atmosphere and in food chains, growing risks of environmental accidents and threats to life support systems.

Although there are existing laws dealing directly or indirectly, with several environmental matters, it is necessary to have a general legislation for environmental protection. Existing laws generally focus on specific types of pollution or on specific categories of hazardous substance. Some major areas of environmental hazards. There are inadequate linkage in handling matters of industrial and environmental safety, control mechanisms to guard against slow, insidious build up of hazardous substances, especially new chemicals in the environment, are weak. Because of a multiplicity of regulatory agencies, there is need for an authority which can assume the lead role for studying, planning and implementing long term requirements of environmental safety and to give direction to, and coordinate a system of speedy and adequate response to emergency situations threatening the environment.

Conclusion

Before closing I quote Words Worth, the nature poet

> "Sweet is the Lore that Nature brings, our meddling intellect
> Misshapes the beauteous forms of things
> We murder to dissect"

How true, we only destroy nature which had once inspired many a poet to write about its unfading glory. Now the poets may have to turn to the grave yards to seek inspiration to write about Nature's fall from glory.

ECOLOGICAL CRISIS IN THE PERSPECTIVE OF CHRISTIAN CONSCIOUSNESS

J.G. Valan Arasu

Introduction

The ecologists are often viewed as persons, who are insensitive to the sufferings, pains and needs of the people, due to their involvement in protecting God's creation from the destruction threatened by human beings. Presently, we tend to be suspicious of the motives behind the activities, which are undertaken by the Ecologists. We ask questions such as why should these ecologists waste their time and energy in studying and conducting research on nature? Why can't they utilize the same energy and resources for the growth and development of mankind that is socially marginalized and does not even have fundamental needs taken care of, such as food, clothing and shelter, etc?

These questions arise because the common people have failed to observe the intrinsic inter-relationship between the earth and human beings. We have failed to recognize the fact that human beings too are part of the same vast ecosystem, and their very survival is at stake. Therefore it is fitting "for ecological prophets" to call out for a global awakening of consciousness which will generate and guarantee a rational loyalty to our planet as a whole, and to its infinite variety of life species and material things in particular"[1], so that we may learn to render respect to every form of life regardless of its worth to us.

Christianity as a Cause for Ecological Crisis

There is a tendency in many of the modern thinkers to place the entire responsibility of the existing ecological crisis on Christianity. They accuse Christianity on the basis of four grounds: theological, the doctrine of salvation, material affluence as the sign of God's blessings and the notion that science has emerged from a Christian religious and social culture. Theological grounds mean that Christianity, through its theological doctrines, directs the hearts of humans to look for a better world in heaven.

The earth on which we live is a temporary place to dwell where one is given license to make use of various objects for oneself and travel towards heaven without contributing anything positively towards the earth. In this sense Christian theology has indirectly granted a license to dominate and exploit creation instead of tending it, as a loving caretaker.[2]

The doctrine of salvation gave undue importance to the salvation of one's soul. The human goal during one's life on earth was to strive after the purification and sanctity of the soul so that the salvation of the soul would be a possibility. Therefore "in order to become a saint, one ought to prefer the spiritual to the material, one has to be detached completely from 'worldly involvements' and strive solely after the 'spiritual pursuits'".[3] In this pursuit of salvation one need not look into the welfare and over all good of our planet. What is important is one's own salvation. Christianity, which is based on the outlook of Biblical spirituality of Jews, considered material affluence a sign of God's blessings[4] and encouraged people to accumulate as much wealth as possible, through whatever means available–exploiting the inexhaustible raw materials, and the cheap labour of the masses.

"The modern science and technology emerged from Christian European culture" accelerated the pursuit of exploitation of the nature as fast as possible.[5] While Lynn condemning Christianity for ecological crisis says, "We shall continue to have a worsening ecological crisis until we reject the Christian axiom... Both our present science and technology are so tinctured with orthodox Christian arrogance towards nature."[6] Mark Spalding, too while reflecting on his Christian life in relation to the environment says, "I made or no connection between my love of, and concern for the environment and my Catholic faith. The two developed quite independently ... because my spiritual direction came from people who, for the most part, rarely made these connections themselves."[7]

Is Christianity Really Responsible for Ecological Crisis?

These thinkers and ecologists hold Christianity responsible for the various ecological degradations caused. It is mainly because they have viewed Christianity from a narrow perspective rather than from a holistic perspective. The focus of Christianity is Jesus Christ and his teachings. His life and teachings are the fundamental foundation for Christianity. Christianity stands and acts on the pillar of Christ's love (agape): an unconditional love for everyone irrespective of one's status in the world. All are brothers and sisters in the Kingdom of Abba Father. A life led on the basis of true 'Agape' will neither exploit nor torture animals and nature for selfish reasons or for fun and enjoyment. The tendency and attitude to dominate the earth without respect for its own laws, needs and rhythms is not due to Christian doctrines, but due to a lack of 'agape'. It is lack of the actualization of Christ's Agape in one's day-to-day life-journey that has resulted in ecological crisis.

Christianity clearly teaches the importance of a just society and the appropriate recognition of human dignity. But in our actual life experience, we come across instances of grave violation of justice and human dignity. In these cases, can we accuse Christianity for the recurring events of violation of human dignity? The answer is that we cannot. So too in the real situation it is not the Christian religion which is responsible for the injustice caused to nature, but the individuals who manipulated and exploited nature, instead of loving and caring it, are responsible for the same. Further, extensive environmental degradation, according to Rene Dubos, "began long before the biblical era; because every civilization abused their environment to some degree by deforestation and over-intensive grazing."[8] Thus, every civilization before the advent of Christianity in one way or the other is accountable for the present world-wide-ecological deterioration.

The Bible's Mission to "Subdue the Earth"

Though the causes for our ecological crisis, criminal exploitation, destruction and deformation of God's creation are many in number, yet a few authors argue that the misunderstanding of the Bible's mission to "subdue the Earth" (Gen 1:28) to a certain extent is an 'Intellectual' cause for our ecological crisis. It was this Biblical charge, says Moltmann Jurgen, "lays the intellectual cause for our ecological crisis: unlimited reproduction, over population of the earth and subjugation of nature."[9] According to Christian Link, it was this incorrect reading of the Biblical mission that led European Christians to render an absolute priority to technology and "a justification for all economic exploitation of nature."[10] But for Cecil J. Saldanha, "Human arrogance, based on a misconception of man's scriptural 'dominion' over the earth, seemed to have set him apart from the rest of creation."[11] Thus, antagonists take Biblical mission to mean 'have vast families and drive other species to extinction'.[12] Before considering these arguments however, it is necessary to understand implications and responsibilities of the Biblical charge given at creation-"be fruitful and multiply and fill the earth and subdue it and have dominion over the fish of the sea and over the birds of the air and over every living thing that moves upon the earth" (Gen 1:28)- from the right perspective in order to protect and to care for the entire creation of God.

The Image and Likeness of God

Gen 1:28 can be better understood in the light of Gen 1:26, that is, God made man in His own "image and likeness". The Hebrew word for the image is '*selem*'. The noun '*Selem*' "predominantly means an actual plastic work, a duplicate, sometimes an idol (I Sam 6:5; Num 33:52: II King 11:18)."[13] According to the Word Biblical Commentary, the most probable opinion for the meaning of "the image" is that "it refers to the mental and spiritual faculties that man shares with his creator"[14] and it is this image of God that distinguishes man from the animal kingdom. This distinction of man from the rest of the creation makes him superior over God's creation and God's representatives on earth.

The notion of being God's representative on the earth because of his God's image is a common oriental view of the king. In the ancient world, "ancient oriental kings were expected to be devoted to the welfare of their subjects, especially the poorest and weaker members of society (Ps 72:12-14). By upholding divine principles of law and justice, rulers promoted peace and prosperity for all their subjects."[15] So too mankind, being made in "the image" of God, has a "commission to rule nature as a benevolent king, acting as God's representative over them and therefore treating them in the same way as God who created them."[16]

Accountability before God

Therefore, the phrase "image of God" should be understood in the perspective of the relational, because the aspect of the relational "describes our unique calling to be in responsible relationship with God, with each other, and with the rest of creation."[17] Being in "God's image" thus demands accountability before God. This accountability means, "On the one hand, the potential to be God's image, thus extending God's presence in the world, and on the other hand, the command to take delegated responsibility for God's creation."[18] It is because of this "God's image" man was appointed as the superior over God's creation, and was placed in the garden with the duty and responsibility of sustenance and watching over creation. But, in exercising this superiority over creatures, man must always "remain subject to the will of God, who imposes limits upon his use and domination over other things."[19] Therefore "the image" of God given to man is a specific "call to imitate in action the one whose image is carried."[20]

The Hebrew verb for "to have dominion" (over the fish of the sea etc.) is *radah*. The meaning of '*radah*' is 'to trample' or to 'press' or 'to tread down.'[21] At the same time the Hebrew word for "subdue" (the earth) is '*kabash*'. '*Kabash*' is part of the same "uncompromising rhetoric within dominion falls: literally it implies trampling under one's feet (cf. Jer 34:11-16)."[22] These meanings indicate to us that there is no doubt with regard to human beings as being appointed by God as superior over the rest of creation. Also these verbs '*radah*' and '*kabash*' establish a hierarchical order between humans and animals and they also connote an absolute subjugation of the creation.[23]

The Concept of Steward

However, the terms 'subdue' and have dominion are balanced and corrected by the other command in the Yahwistic tradition, "to cultivate and care for" the earth, which leads us to view man, as exercising of dominion over God's creation, as the "manager" or "steward", or much better still "the faithful caretaker" or "shepherd."[24] It is wrong however to perceive the verb '*radah*' as a permission for the exploitation or destruction of God's creation.

The expression '*radah*' "has its origin in the language of royal court in Babylon and Egypt.[25] Also the frequent usage of this verb '*radah*' in connection with kingship and royal heritage convince us that it was part of the technical language of royal rule. The royal rule is neither a domination of exploitation, nor a domination of an arbitrary rule.[26] The power of domination should be "Exercised in the cause of right and justice and equality, the way that God exercised it in Israel's history and the way that he intended that man should exercise it as his surrogate. This means, in modern terms, a rational, sensible, humane, intelligent, and thoughtful ordering of God's ordered world, dispensing correctives where they are necessary and furthering propriety where it is clearly proper."[27]

The phrases 'having dominion' or 'subduing' in relation to man being made in God's 'image and likeness' implies that dominion over God's creation does not mean that man can exploit the entire creation of God for his own self-glory and self-development and self-interests. The "humans' dominion over the earth cannot be an absolute power or freedom to use the things of the earth as they please."[28] To "subdue the earth" however means to take an adequate responsibility to care for, to safeguard and to develop God's creation according to the eternal plan, which God has for his own creation. Human dominion means to maintain and to take care of the welfare of God's creation for its own sake. Like a good shepherd he takes care of his flock even at the cost of his very life. Human dominion thus "should be exercised in such a way *as to serve and preserve* the beasts, the trees and the earth itself - all of which is being ruled."[29]

Christianity has produced saints like Benedict, Francis of Assisi, Hildegare of Bigen and Anthony and many others, who through their way of life exhibited and expanded Christ's Agape to the entire creation. St. Anthony (250-356), speaking on the treasure of Gods' message that is contained in the nature says, "My book is the nature of created things and it is always at hand when I wish to read the words of God." So how can we accuse Christianity when it has produced saints who have fallen in love with the nature? In concrete circumstances it is not Christianity as a religion that is responsible; rather each individual irrespective of their religion is responsible for the ecological crisis.

Human Attitudes

In every person's life, his surroundings, circumstances, social background, language and friends play a dominant role in determining the character and attitude of persons. One is influenced, guided and to a certain extent manipulated by ideologies and opinions, which are external to one, but having their influence on the processes of self-formation of internal attitudes. In some the influence may be greater while in some others it may be negligible. The impact and the thought patterns of different thinkers in various period of history indicate that they have successfully contributed to the care and the growth of the whole of God's creation. But a few others influenced people negatively in their hearts and minds, leading them to form thought patterns, which are detrimental to both the animate and the inanimate creation of God.

Jesus said to his disciples, "Nothing that goes into someone from outside can make a person unclean... It is what comes out of someone that makes that person unclean. For it is from within, from the heart, that evil intention emerges" (MK 7:16-21). It is the interior attitudes and inner human frameworks, and the disequilibria within each individual and humanity as a whole that caused the ecological crisis and not a specific religion or ideology. It is correct therefore to search out a few of the interior human attitudes, which urged people to perform certain types of action, which are harmful and exploitative in relation to the entire cosmic organism.

Concomitance of Sin

Sin is essentially the 'rejection of God' and 'His divine plan'. Sin is a refusal to acknowledge God as our source. As a result, humans upset the cordial relation, which they should have with themselves, with others, as well as with all creatures.[30] Sin, according to the Bible, is not only "distorts inter-human relations and human-Divine relations, it also affects the life-sustaining harmony between humans and the earth."[31]

The inter-relation between human sin and ecological destruction, as well as the net impact of human sin on the earth, can be verified right from the beginning of human history that is described in the Bible. The Bible constantly sees and describes the moral disorder of humans as the cause for cosmic disorder. Therefore, whenever humans' corrupt practices and sins increased, God decided to destroy not only the humans but also the other forms of life on earth, in order to bring back a new cosmic order based on cordial and complementary relationship with every creation. The correlation between human sins and ecological destruction is cited not only in the book of Genesis but rather this correlation is flowing through out the scripture.

In the fall-story, the soil was accursed because of human disobedience of the commands of Yahweh. The Bible says, "To the man he said, because you listened to the voice of your wife and ate from the tree of which I had forbidden you to eat, accursed to be the soil because of you!" (Gen 3:17-19). In the flood account God said to Noah after witnessing the actual corrupt practices, "I have decided that the end has come for all living things, for the earth is full of lawlessness because of human beings. So I am about to destroy them and the earth" (Gen 6:13). As a result of lawlessness of human beings "all living things that stirred on earth perished; birds, cattle, wild animals, all the creatures swarming over the earth, and all beings. Everything with the least breath of life in its nostrils, everything on dry land, died. Everything on the face of the earth was wiped out, people, animals, creeping things and birds; they were wiped off the earth" (Gen 6:21-24). Also, the cities of Sodom and Gomorrah were destroyed by sending brimstone and fire because "their sin is so grave" (Gen 18:21).

The prophetic tradition, too, specially brings out clearly the intrinsic connection between human disobedience of Yahweh's covenant, and the resulting destruction of creatures. Prophet Hosea says, "wild animals also, and birds of the sky, even the fish in the sea, will disappear", because there is no loyalty, no faithful love and no knowledge of God in the hearts of Israelites. Only perjury and lying, murder, theft, adultery and violence, bloodshed after bloodshed are found in the mind of Israelites. (Hos 4:1-3). Prophet Isaiah's writings too clearly bring out the correlation between human sin and destruction of earth. According to Isaiah, "Ravaged, ravaged the earth will be, despoiled, despoiled, for Yahweh has uttered this word. ... The earth is defiled by the feet of its inhabitants, for they have transgressed the laws, violated the decree, broken the everlasting covenant. That is why the curse has consumed the earth and its inhabitants pay the penalty that is why the inhabitants of the earth have been burnt up" (Is 24:5-6).

Jeremiah too is of the view that, "the animals and birds are dying as a result of the wickedness of the inhabitants" (Jer 12: 4). Prophet Zephaniah pronounced God's Judgment on all creation since humans broke God's covenant. "I shall sweep away everything off the face of the earth, declares Yahweh. I shall sweep away humans and animals, the birds of the air and the fish of the sea" because of Judah's sins. For they have turned their back on Yahweh who do not seek Yahweh. (Zep 1:1-4). It is interesting to note that whenever man sinned, it is not only man alone who repented for his sins rather the creatures of the earth participated completely in the penance undertaken by human race. Cattle and sheep of Nineveh fasted in solidarity with the human to atone for the city's wickedness. It is written, "No person or animal, herd or flock, may eat anything; they may not graze, they may not drink any water. All must put on sackcloth and call on God with all their might; and let everyone change his mind and relent and renounce his burning wrath, so that we shall not perish" (Jon 3:7-9).

It is not only the Biblical incidents that prove the close concomitance of sin and ecological destruction, but contemporary events too prove the concomitance of sin and the destruction of God's creation. The two world wars, a number of civil wars, the concentration camps of the Nazi caste conflicts, increasing number of earth-quakes, material developments at the cost of ecological degradation, the multiplicity of injustices, the violation of human dignity and the alarming ecological crisis are the net results of twenty-first century humans' refusal to accept their dependence and render their obedience to God.

The ecological crisis of which we are aware today is the net result of the happenings of the last thirty years. It is during this period of history that half of the total quantities of coal, number of mineral deposits and crude oil reserves have been extracted. It is said, "Between the year 1850 and 1950 one species disappeared per year; but around 1990 ten species are disappearing per day."[32] It shows the existence of proximate relationship between human sin and ecological crisis. Pointing out the correlation between human sin and ecological crisis Pope John Paul II rightly said human sin against God "resulted... in the earth's "rebellion" against him.....when man turns back on the Creators' plan, man provokes a disorder which has inevitable repercussions on the rest created order"[33] "Just as faithfulness to the covenant...brings about order and harmony between people and in the cosmos itself, sin destroys human [harmony] and brings about cosmic chaos."[34] Therefore sin is the ultimate polluter of the world, because "due to sin, earth and human family suffered from disruption of the harmony that God placed in creation, a harmony dependent upon each fulfilling its role as complementary creature to one another.[35]

Banishment of God from the World

Another major cause for the ecological crisis is the banishment of the concept of God from the hearts of humans. Since "the modern age has banished God from the world ... it has therefore banished beauty. Everything has become 'profane', exiled from the sphere of the 'holy' and therefore everything has lost its meaning."[36] The failure of humans to see, to experience and read "God's Excellency, His wisdom, His purity and love, [as appearing] in every thing: sun, moon and stars"[37] made them idolize and "honour the work of [their] hands rather than works that come more directly from the hand of God."[38] The book of Genesis presents God as driving the humans from the garden. However, today, the humans have driven God - author of all creation, including the world–away from the world and "have occupied even the most beautiful places in the name of science, technology, tourism and defense."[39] As a result, "technology became god, and the high priest of this newly founded "religion" scarified the riches of this land–all the living organisms of the cosmos–to that god of his own making."[40]

This led to the exploitation and devastation of God's earth, the abuse of God's creation, and pleasure and happiness were derived from the infliction of pain on humans, animals, plants and environment. Therefore, the ecological crisis which humans experience is a rebellion on the part of nature caused by their refusal to carry out their role as cooperators with

God in the work of creation and thus trying to acquire a place for themselves which is due to God alone.[41]

Loss of Human Respect

Another cause for the ecological devastation is the loss of mutual respect amongst humans, and the confusion and disharmony created by it in the sphere of human relationships. The lack of acceptance and respect among humans led them to consider the workers "not as the 'subject' but only as the 'object' of production."[42] Not human values and relationships but economic interests and production have taken priority over the good of individuals and the people at large. People are valued, not for what they are in themselves, but for what they own and contribute towards the production of material goods. Human beings are no longer considered as beings made in the image and likeness of God rather they are considered as mere living technical machines which can produce material products. The producers and managers as a result are interested in increasing the efficiency of the human machinery that is interested in the extraction of human labour, energy and efficiency to make large profit for self and not interested in improving the working conditions of the people at large.

However "the dignity of the human person is the ultimate guiding norm for any sound economic, industrial or scientific progress"[43] has been forgotten for ever. The human persons were no more respected and regarded as human persons with dignity rather persons were "turned into a commodity of calculated market value."[44] In this environment it is most likely that we lose our respect and regard for other creatures, because "the stance men and women take towards their fellow human beings is of a piece with their stance towards the earth."[45] When humans have no regard for their own species, it is unlikely that they will have regard for the creatures. The actual experience of using more and more of the earth's resources and extinguishing thousands of species of plants and animals, endangering hundreds of others, tells us of our distorted human relationships and a lack of respect for life. From the above stated reasons one can straightforwardly conclude that the destruction of the environment is the result of an unnatural and reductionist vision of human dignity leading to the practice of human dignity in proportion to the wealth one possesses during his/her life time on this earth.

Human Dignity in Proportion to Wealth

In our day-to-day situation there is a deep-rooted tendency in the heart of every person to measure personal worth of a person by possessions he has accumulated during this life. Over and above the value of the person is measured according to the things he or she possesses and not on the moral worth and character of the person.[46] As a result significant respect and honour is given to a person who has accumulated sufficient wealth for the generations to come. The rich are well respected in all social gatherings, and their words and advice carry more weight than those of the poor, and thus often they gain supremacy over others who do not have sufficient wealth. This undue and unquestioned respect given to the rich, irrespective of the means applied in their pursuit to accumulate wealth induces people to think that what matters is not "how", but "how much", wealth one has succeeded in accumulating.

It is also a fact that in the secret heart of every person there is a deep longing for self-respect, self-worth or recognition. Children at even a very young age, wish to become doctors, or rich people, or engineers etc, and never wish to be ordinary or daily wage earners. It is because in their day-to-day experience they witness that only learned, knowledgeable and wealthy people are well respected and honoured within the context of human society. This sort of experience creates an urge in the children, at a young age, to dream of and imitate the socially well-respected people in their day-to-day living.

Irrespective of their needs, the attitude of considering the possession of wealth as the source for granting self-repudiate urges people to be greedy and to accumulate as much wealth as possible. It influences people to work to achieve a high level of material affluence without counting the cost of the harm done to the natural environment that supports us. It creates a growing frustration of what to seek and achieve in life. All these innovative and wish-able ideas allow humans finally to end up in destroying numerous forms of life, and disturbing the balanced cycles of Nature in order to build a bright home, satisfy their selfish interests and gain self-respect for them.

Greed is defined as "the inordinate desire for material goods in the form of possessions"[47] Christ includes greed among the evil things which come forth from the heart (Mk: 7:22). St. Paul considers this greed as the root cause for all the evils: "love of money is the root of all evils" (I Tim 6:6-10). The sinfulness of greed consists in not devoting "the goods of this world to their proper purpose."[48] Whenever humans possess and accumulate wealth, without meeting these demands, they are directly involved in abusing God's goods. For "every aspect of our ecological crisis reveals selfish greed, unbridled desires to possess, control, and exploit for one-self what is meant to be for the service of all."[49] The selfish greed made us to use our "good technological abilities to exploit God's creation, to destroy the balance of nature, and to deform ...In fact it stands in danger of conflagration, in the face of nuclear war."[50]

Greed has also infused an urge within humans for greater profit, which in turn induced the planners, procedures, entrepreneurs and industrialists to take short cuts and promote developmental patterns, which are environmentally disastrous.[51] Humans' greed for material possessions has also blinded them to the pain and suffering which they have caused to

Mother–earth by their quest for natural resources. As a result, "the rich resources of water, atmosphere, and soil, as well as of plant and animal products, oceans and shores, fall victim to human greed."[52] The reason for this selfish greed is the misunderstanding of the earth–realities and the taking of the created world for the ultimate reality: "to cling to veil instead of the divine mystery". This partial view of the earth allowed humans to use our mother–earth as a "storeroom from which we take whatever we find useful to fulfill our needs"[53] and for short term comforts and luxuries.

Loss of the Sense of Solidarity

The practice of assessing human worth in proportion to accumulated wealth and the mentality of greed rooted in human psyche has resulted in the loss of a sense of solidarity with other creatures. Further, this has led to ignorance of the wholeness of nature and of mutual interdependence and the earth as a living whole. Human beings have failed to remember that they are one species among many species and a part of the rest of God's creation. They are not apart from other creatures rather they depend on other creatures for their existence and survival.

It is this sense of solidarity, which brings all beings together and makes them feel concerned about one another, and binds them with mutual obligations and respect. For "the welfare of the individual and the common weal depend on each other reciprocally."[54] Today the individual has become more conscious of his dependence upon society. But he does not consider this dependence as a positive asset, as an organic tie, as a protective force, but rather as a threat to his natural rights. Man more and more considers himself as a solitary being and attempts to protect his own existence and that of those who are closest to him, to satisfy his personal desires and to develop his innate abilities. In this process he forgets the fact that it is nature which provides man with food, clothing, a home, and the tools of work etc.[55] He has forgotten the fact that his life is made possible through the labour and the accomplishments of the many millions past and present who are all hidden behind the small word 'World'. When this sense of dependence is lost, it leads to "individualism, in which one claims his own rights without wishing to be answerable for the common good,"[56] leading to "selfish greed, unbridled desires to posses, control, exploit for one self what is meant to be for the service of all,"[57] and eventually culminates in ecological disaster.

The Eco-system is like a spider's web: "all strands of the spider's web are inter-connected … one cannot break, remove, add or put a stress upon any without necessitating an adjustment throughout the entire web."[58] Ignorance and inadequate understanding of the natural world, its laws, and interrelatedness make us interfere "in one area of the ecosystem without paying due attention both to the consequences of such interference in other areas and the well-being of future generations."[59] Motivated by self gains, man continues to work on establishment of industries after industries, without giving significant thought to the effects like pollution, greenhouse effect, unrestricted deforestation, burned fossil fuels and disposed of industrial wastes, have a detrimental influence on environment. These cause atmosphere pollution and environmental degradation for the generation to come.

Science at large has conducted enormous tests and experiments–nuclear tests, experiments on animate and inanimate beings, etc.- without being in a position to assess the biological and ecological disturbances that could result from a particular test. For example, "no place on Earth can be shown to be safe for disposing of nuclear waste, because once the radioactive material is absorbed into the biological cycle, it contaminates everything."[60] Yet all the nations of this world have joined in the process of registering themselves in the race of becoming nuclear powers in the world.

Conclusion

Natural law therefore demands that humans, who are a family member in God's creation, "must live our life responsibly in harmony with its other parts, while fulfilling our role as trustees of the earth."[61] For "it was the Creator's will that man should communicate with nature as an intelligent and noble 'master' and 'guardian' and not as heedless 'exploiter' and 'destroyer'".[62] Pope John Paul II thus invites entire human race to live and exercise the quality of solidarity with the rest of creation successfully because according to him "only that virtue allows one to take full co-responsibility for the fate of the world and the other person..."[63]

The causes for the ecological crisis attributed here clearly indicate that it is not simply an automatic effect of the cycle of energy. None other than human beings rather ultimately causes it. It results from their behaviour patterns and has its origin within the interior framework, interior attitudes and thought process of man. As a result science, technology, industrial developments and other possible developmental growths "could do nothing apart from the act of human beings" and no human person can be exempted from the responsibility of the ecological crisis caused. It is clear; therefore, since we are doing it, only we can stop it if we want. St. Francis of Assisi became a lover of nature and reformer only after "a long and demanding novitiate"[64] in which he lived with the poor and lepers. It was at the end, not at the beginning of his life that Francis of Assisi composed the hymn to the Sun and considered the various elements of creation as brother and sister creatures. Ecological crisis thus cannot be solved by a few definitive technological restraints like closing down a few industries and stopping certain experiments. It can be only be solved by a radical conversion and transformation of human attitudes, perspectives and the basic values we entertain about nature. For reverence and respect are rooted in the severest sort of asceticism alone.

Every civilization, history, cultural heritage and event has a unique message to reveal to the entire human community. So too, the alarming ecological crisis that exists today has a distinct and a definitive message to convey to the entire humanity. The message is that human beings cannot take the earth and creation of God for granted; rather we have an untransferable and undeviating obligation to the surroundings, which we ought to realize. Therefore, what is expected of us is to perform those actions, which are genuinely oriented towards the protection of our environment.

Notes

1. R.J. Raja, "Eco-Spirituality in the Psalm", *Vidyajyoti Journal of Theological Reflection,* Vol. LIII, December 1989, pp. 637-638.
2. Charles Cummings, Eco-Spirituality: Toward a Reverent Life, New York, Paulist Press, 1991, Pp. 102-103.
3. A. Pushprajan, Ecological World-View for a Just Society, New Delhi, I.S.PC.K. Publication, 1992, pp. 12-13.
4. J.B. Chethimattam, "*A Philosophical Approach to the Ecological Crisis*", Journal of Dharma, Vol. XX, Jan-March 1995, pp. 18-19.
5. Sean Mcdonagh, To Care for the Earth: A Call to a New Theology, New Mexico, Bear and Company, 1987, p. 137.
6. As quoted by Ibid., p. 137.
7. Mark Spalding, "Earth Crisis", *Priest and People*, Vol. IX No. 2, p. 62.
8. As quoted by Sean Mcdonagh, op.cit., p. 137.
9. Jurgen Moltmann, *God in Creation; An Ecological Doctrine of Creation*, London, SCM Press Ltd., 1985, p. 149.
10. Christian Link, "Ecological Crisis and Christian Ethics", Theological Digest Vol. XXXI, 1984, p. 149.
11. Cecil Salanha, "Management of the Earth: A Responsible Response," *Journal of Dharma,* Vol. XX, Jan-March 1995, p. 15.
12. Mark Spalding, art.Cit., p. 65.
13. Gerhard Von Rad, *Genesis a Commentary*, London, SCM Press Ltd, 1972, p. 60.
14. Gordon J. Wenham, Word Biblical Commentary, Genesis 1-15, Vol. I, Texas, Word Books Publisher, p. 30.
15. *Ibid.*, p. 33.
16. *Ibid.*, p. 33.
17. Loren Wilkinson ed., Earth-keeping in the Nineties: Stewardship of Creation, Michigan, William B. Eerdmans Publishing Company, 1991, p. 285.
18. Mary Evelyn Jegen ed., *The Earth is the Lord's: Essays on Stewardship*, New York, Paulist Press, 1978, p. 24.

19. Peter C. Phan, "Pope John Paul II and the Ecological Crisis," The Irish Theological Quarterly, Vol. LX, No. 1, 1994, p. 61.
20. Claus Westermann, Genesis 1-11, a Commentary, London, SPCK, p. 155.
21. Loren Wilkinson ed., *op cit.*, p. 285.
22. Bruce Vawter, On Genesis, A New Reading, New York, Doubleday and Company, 1977, p. 60.
23. Claus Westermann, *op cit.*, p. 158.
24. Peter C. Phan, art.cit., p. 68.
25. Claus Westermann, *op cit.*, p. 158.
26. Bruce Vawter, *op cit.*, p. 58.
27. *Ibid.*, p. 59.
28. Peter C. Phan, art. cit., p. 6.
29. Loren Wilkinson ed., *op cit.*, p. 287.
30. Thomas Pazhayampallil, Pastoral Guide, Vol. I, Bangalore, K.J.C. Publications, 1984, p. 190.
31. Sean Mcdonagh, *op cit.*, p. 125.
32. Johan Van Klinken, "'The Integrity of Creation': The Third Point of the JPIC Process. Ecology between Theology and Science", *Concilium,* No. 4, 1991, p. 67.
33. John Paul II, "Peace with God the Creator, Peace with all of Creation", *L'Osservatore Romano,* December 18, 1989, No. 3 and 5, p. 1.
34. Sean Mcdonagh, *op cit.*, pp. 125-126.
35. John Hart, *The Spirit of the Earth,* New York, Paulist Press, 1984, p. 157.
36. John Kulandai, "Basics in Eco-Spirituality", *Vidyajyoti Journal of Theological Reflection,* Vol. LVI, September 1992, p. 496.
37. Charles Cummings, *op cit.*, p. 54.
38. *Ibid.*, p. 93.
39. John Kulandai, art. cit., p. 496.
40. Ibid., p. 496.
41. John Paul II, *Centesimus Annus*, Bombay, St. Paul Publications, 1991, No. 37. p. 73.
42. Sean Mcdonagh, *op cit.*, p. 74.
43. John Paul II, art. cit., *L'Osservatore Romano,* No. 7, p. 1.
44. A. Pushparajan, *op cit.*, p. 70.
45. Charles Cummings, *op cit.*, p. 103.
46. Thomas Pazhayampallil, *op cit.*, p. 784.
47. Karl H. Peschke, *Christian Ethics: Moral Theology in the Light of Vatican II,* Vol. II, Bangalore, Theological Publication in India, 1992, p. 693.
48. *Ibid.*, 694.
49. J. Neuner, "Mother Earth; Jun. 5th World Environment Day", *Vidyajyoti Journal of Theological Reflection,* Vol. LVIII, May 1994, p. 321.

50. Limouris G. ed., *Justice, Peace and the Integrity of Creation Insights from Orthodoxy,* Geneve, Wcc, 1990, p. 15, as quoted by R.J. Raja, "As it was in the Beginning", *Vidyajyoti Journal of Theological Reflection, Vol. LV, 1991, p. 683.*
51. Thomas Manninzhath, art. cit., p. 15.
52. Pawel Goralczyk, "The Formation of an Attitude of Ecological Responsibility", *Communio,* Vol. XXI, No. 3, 1994, p. 542.
53. J. Neuner, art. cit., p. 321.
54. Karl H. Peschke, *op cit.,* p. 521.
55. Ahindra Ghosh, "Current Ethical Challenges–Some Reflections", T.D. Singh and Samaresh Bandyopadhyay, ed., Thoughts on Synthesis of Science and Religion, Calcutta, The Bhaktivedanta Institute, 2001, p. 223.
56. Ibid., p. 522.
57. J. Neuner, art. cit., p. 321.
58. Thomas Manninezhath, "A Vedantic Response to the Ecological Crisis", *Journal of Dharma,* Vol. XX, Jan- March 1995, p. 26.
59. John Paul II, art. cit., *L'Osservatore Romano,* No. 6. p. 1.
60. Sean Mcdonagh, *op cit.,* p. 100.
61. John Hart, *op cit.,* p. 163.
62. John Paul II, *Redemptor Hominis,* No. 15, as found in Michael Walsh and Brian Davies eds., *Proclaiming Justice and Peace,* Theological Publications in India, p. 254.
63. Pawel Goralczyk, art. cit., p.549. Cf. John Paul II, *Sollicitudo Rei Socialis,* Bombay, St. Paul Publications, 1990, No. 38.
64. Drew Christiansen, "Ecology, Justice and Development", Theological Studies, Vol. LI, 1990, p. 75.

Bibliography

Chethimattam, J.B. (1995), "A Philosophical Approach to the Ecological Crisis". *Journal of Dharma:* Vol. XX, Jan-March.

Christiansen, Drew (1990), "Ecology, Justice and Development". Theological Studies: Vol. LI, 1990.

Cummings, Charles: Eco-Spirituality: Toward a Reverent Life. New York: Paulist Press, 1991.

Ghosh, Ahindra (2001), "Current Ethical Challenges - Some Reflections". T.D. Singh and Samaresh Bandyopadhyay, ed.: *Thoughts on Synthesis of Science and Religion*. Calcutta: The Bhaktivedanta Institute.

Goralczyk, Pawel (1994), "The Formation of an Attitude of Ecological Responsibility" *Communio*: Vol. XXI, No. 3, 1994.

Hart, John (1984), *The Spirit of the Earth*. New York: Paulist Press.

Jegen, Mary Evelyn ed. (1978), *The Earth is the Lord's, essays on stewardship*. New York: Paulist Press.

Kulandai, John (1992), "Basics in Eco-Spirituality". *Vidyajyoti Journal of Theological Reflection*: Vol. LVI, September.

Limouris G. ed. (1990), *Justice, Peace and the Integrity of Creation Insights from Orthodoxy*. Geneve: WCC.

Link, Christian (1984), "Ecological Crisis and Christian Ethics". *Theological Digest:* Vol. XXXI.

Manninezhath, Thomas (1995), "A Vedantic Response to the Ecological Crisis". *Journal of Dharma*: Vol. XX, Jan-March 1995.

Mcdonagh, Sean (1987), To Care for the Earth: A Call to a New Theology. New Mexico: Bear and Company.

Moltmann, Jurgen (1985), *God in Creation; An Ecological Doctrine of Creation*. London: SCM Press Ltd.

Neuner, J. (1984), "Mother Earth; Jun. 5th World Environment Day". *Vidyajyoti Journal of Theological Reflection*: Vol. LVIII, May 1994.

Paul II, John (1989), "Peace with God the Creator, Peace with all of Creation". *L'OsservatoreRomano*: December 18, 1989, No. 3 and 5.

Paul II, John (1991), *Centesimus Annus*. Bombay: St. Paul Publications, No. 37.

Paul II, John, *Redemptor Hominis. Bombay: St. Paul Publication,* No. 15.

Paul II, John (1990), *Sollicitudo Rei Socialis*. Bombay: St. Paul Publications, No. 38.

Pazhayampallil, Thomas (1984), Pastoral Guide, Vol. I. Bangalore: K.J.C. Publications.

Peschke, Karl H. (1992), *Christian Ethics: Moral Theology in the Light of Vatican II* Vol. II. Bangalore: Theological Publication in India.

Phan, Peter C. (1994), "Pope John Paul II and the Ecological Crisis". Vol. LX, No. 1.

Pushprajan, A. (1992), Ecological World-View for a Just Society. New Delhi: I.S.PC.K. Publication.

Raja, R.J. (1989),"Eco-Spirituality in the Psalm". *Vidyajyoti Journal of Theological Reflection:* Vol LIII, December.

Raja, R.J. (1991), "As it was in the Beginning". *Vidyajyoti Journal of Theological Reflection*: Vol. LV.

Salanha, Cecil (1995), "Management of the Earth: A Responsible Response". Journal of Dharma: Vol. XX, Jan-March.

Spalding, Mark, "Earth Crisis". *Priest and People:* Vol. IX No. 2.

Van Klinken, Johan (1991), "The Integrity of Creation: The Third Point of the JPIC Process. Ecology between Theology and Science". *Concilium*: No. 4.

Vawter, Bruce (1977) On Genesis, A New Reading. New York: Doubleday and Company.

Von Rad (1972), Gerhard: *Genesis a Commentary*. London: SCM Press Ltd.

Walsh, Michael and Brian Davies eds.: *Proclaiming Justice and Peace*. Bangalore: Theological Publications in India.

Wenham, Gordon J., Word Biblical Commentary, Genesis 1-15. Vol. I, Texas: Word Books Publisher.

Westermann, Claus: Genesis 1-11, a Commentary. London: SPCK.

Wilkinson, Loren ed. (1991), Earth-keeping in the Nineties: Stewardship of Creation. Michigan: William B. Eerdmans Publishing Company.

LEGAL ASPECTS OF ECOLOGY AND HINDUISM

Mona Purohit

Nothing is new about the correlation between Religion and Ecology. All Religions have aimed their policies towards ecological balance so also it is with Hinduism. No body can deny this coexistence. The importance of Ecology is well recognized by all the religions of the world. The only demand of time is that the Law Makers and Interpreters should take active steps towards adopting Natural Law Theories, which means recognition of moral or ethical duties of man towards Ecology. Before pondering into the legal aspect, I would like to discuss the concept of Hinduism with special reference to Ecological Balance in the global scenario.

Comprehensive life and recognition of any thing or when any material thing has its own entity world wide, then it is called "Globalization". The word "Ecology" has a broad meaning that includes the correlation between organism and their environment. It deals with the spacing of people and institutions and the resulting interdependency. According to U.N. Council on Environmental Quality, "Ecology is the Science of the intricate web of relationship between living organism and their living and non-living surroundings". "Religion" is a matter of conscience and it is one of the various systems of faith. It concerns over what exist beyond the visible world.

It is purely made by beliefs upon "God", who has created this Universe and in present scenario, it has its own entity of faith like *Hinduism, Islam, Buddhism, Jainism, Christianity*, etc. It is the base of devotion and commitment of right things and concepts. It is based on the concept of super natural power, which is above and beyond the natural and unexplained by natural laws or phenomena. Basically it is the direct influence or action of God on earthly affairs. This is the right and positive concept of Religion. But, when we take the old approach by establishing the relationship between "Ecology and Religion", then the negative concept can be found out which was hidden. According to this concept, Religion is superstition. It means all the beliefs and notions are not based on reason and knowledge. It is irrational, unknown and mysterious. It may be a blindly accepted belief or notion also. But when we honestly, without any bias, with our devotion, try to make a global bond between "Ecology and Religion", it does not result in a sense of superstition.

The concept of Religion is not new but very old from time immemorial. Even in early days it was developed and originated by the "Super Natural Power". The Sun, the Moon, the Earth, the Sky, Fire, Air, etc., all come within the concept of Religion. Religion is the primitive form of Ecology. In a simple sense, it is a tool to control the mass and maintain balance between the organs of Ecology. Sixteen sacraments of our life are the best example of this concept. Marriage is a holy sacrament in Hindu Religion, when a person comes in this world he has to add up something to this world and for the fulfillment of this requirement, marriage is mandatory. On the other hand, death as the last sacrament of Hinduism is necessary to maintain a balance in environment. When a person belonging to Hindu Religion dies, her/his body comes within the touch of fire to burn him completely to save our atmosphere from pollution. In the same way, when a Mohammedan dies, his/her body is buried in the depth of Earth. According to their belief, that body dissolves through biological actions under the layer of the earth and does not come with the contact of environment or atmosphere. These are the methods to save our atmosphere from pollution and maintain a balance among the organs of Ecology. So, Religion is a source to bond and unite the Ecological World.

In old and primitive age, man was dependent on nature and there was nothing like Religion and beliefs. The prime instincts of those people were hunger and sex. The growth of culture and civilization has started a new era. Since then, people felt the necessity of ecosystem in every moment of life. Even the Air we breathe and Water we drink, are found in Ecological World (both are essential for life). Balance in the Ecology is like equilibrium in nature. After the birth of civilization, people tried to affect the Ecology adversely, which resulted in danger. His use of natural resources like Water, Air, Land, food, etc. affected the dynamics of plants and animals life on Earth by altering ecological balances. Because, all activities of a person's daily life, from morning to evening are governed by various positions of the Sun and the Moon. All the system of our life is conserved with the concept of purity.

According to ancient Indian mythology, God prevails in the whole Universe. This concept has helped Indian to maintain a constant link with nature. Even the concept of five elements viz: Air, Water, Fire, the Earth and the Space are directly concerned with our Ecology. Therefore, it can be said that all the values and norms originated by Religion have a great impact upon Ecology. The whole concept of Religion is interconnected with the concept of Ecology. Four *Ashramas* (Various stages) of life are the best example of it, in which the time of life of *Vanprastha* and *Sanyas Ashramas* were completely carried in the forest and directly dependent on the forest. It shows an interrelation between Religion and Ecology in a global scene. Even plants and trees like *Neem, Tulsi, Bargada,* etc. have a great importance and have profound contribution in the life of a human being. It shows a very close relationship between Religion and Ecology.

We can easily understand the concept of Ecology and Religion by knowing the interrelation between Religion and Environmental Protection:

"Religion commonly pertains to the spiritual domain while science and technology are viewed as systems pertaining to the material domain. Issues of material development and progress has, therefore, been left to the charge of scientific experts while religious belief's has been left to guide the inner spiritual development of the individuals. The ecological crises in the ultimate analysis can lead to the under mining of very basis of material existence."[1]

It is very clear that Indian religious and cultural traditions are very rich and diverse in values, and play a vital role in protection of environment. Ecology gives a man a lot of opportunities to make himself wise and responsible. It is also said that the God is omnipresent in the manifestations of nature.

"Hindu Religion is one of the oldest Religions of the world."[2] To establish harmony with nature is the prime object of everyone since V*edic* time. So it is the sacred duty of mankind to protect our Ecology and Environment. We can feel respect and intimacy for Nature and Environment in Hindu Religion and its scriptures. Hinduism expected the belief upon natural organs like Space, the Earth, Air, Water, Fire, etc. and recognized them as creations of God.

But felling of trees, pollution of Air, Water and Land was regarded as a sin and maintaining purity with them is the duty of every one. Ecological elements have religious sanctity for human life and are regarded as indispensable in the life of human beings. Trees are the most important organs of the Ecology according to Hindu mythology and they are worshiped as *Vriksha Devta* (Tree Deity) "Planting of trees was also treated as scared religious duty and work of great virtue and regarded one tree to equal ten sons."[3]

Even *Yagnavalka Smrity* stated about the obligation of men to protect plants and animals for his own survival and prosperity.

Due to technological development the clean atmosphere of earth was heavily polluted. The most important cause of Ecological imbalance is materialism. Non-possession of things can support in this direction. It means simple living might serve a model for a sustainable development. We need to learn from our culture, which has not taken a modern path of development. In short, the most significant reason for environmental degradation is deteriorating relationship between men and environment because of rapid rate of exploitation of natural resources, technological development and industrial expansion. Human activities at the wide level are responsible for that. Deforestation is also playing a valuable impact on environment imbalance. In fact, the urbanization and culture of developed countries are responsible for this problem. Poverty is also playing a vital role in this direction. Due to poverty people exploit the natural resources excessively to fulfill there basic needs like food for life. So, it is necessary that responsible steps should be taken to bring the poor people above the poverty line.

In the same field "The international forum on Religion and Ecology" is playing an important role. This forum recognizes that Ecology need bear dialogue with other disciplines i.e. science, ethics, economics, education, public policy, etc. in seeking comprehensive solution to both global and local environmental problems."[4]

So, to solve this problem we should encourage a well-controlled and well-managed process of urbanization in order to curb rural urban migration and other problems. The population growth should also be reduced through some comprehensive and effective legislation. There should be restriction on deforestation also.

Today it is well established that pollution of environment is not only a crime but also a tort. In many civilized countries including India, there is a Constitutional duty to protect the environment from pollution. By this type of Law man has been obliged to do something. This is after Austin's analysis of Law, i.e., Positive Law or Command of Sovereign or Kelson's Pure Theory of Law. But before that there was Natural Law period.

"Natural Law as conceived by its exponents is that the Law, which is inherent in the Nature of man or society."[5] The philosophy behind that law is Ethics, Morality, Ideal, Social Justice are the factors by which Natural laws derive its validity. Human conducts are governed by reason or basic human nature.

The same concept now exists in different name like "the concept of the Rule of law in England, the due process of law in the United States and the procedure established by law embodying justice traditions of Natural Law Philosophy wherein the Sovereign body must act under the law in administration of justice. This is in essence of a Natural Law thinking by which the Positive Law is subjected to the inhibition of a moral order."[6]

India being a Welfare State is bound by its duty to provide Political, Social or Economical Justice to its policy. Chapter III, Fundamental Rights and Chapter IV embodying the Socio Economic goal of state policy for welfare of the people, are chapters that enshrined the philosophy of Natural Law.

In the end, one can say that Ecology and Religion particularly Hinduism are so interlinked that in modern time states can use "Natural Law" as an instrument of legal development and legal reform or "Jus civile" can get its validity from "Jus Naturale". We need not borrow any thing from outside because it is inherent in our culture itself.

Notes

1. Khan, I.A. *Environmental Law* Allahabad: Central Law Agency, (2000) 23.
2. Khan I.A. (2000) 24.
3. Khan I.A. (2000) 25.
4. <http:/www.environmet.harword.edu/Religion/main.html.com>
5. Dhyani, S.N., *Jurisprudence and Indian Legal Theory*, Allahabad: Central Law Agency (1999) 35.
6. Dhyani, S.N. (1999) 3.

INTEGRATING ECOLOGY WITH RELIGION

B.N. Behera & P.K. Patra

Maria Montessori, a contemporary thinker reinforced the dictum that nothing can exist in isolation: "The stars, stones and life of all kinds, form a whole in relation to each other and so close in this relationship that we cannot understand a stone without some understanding of the great sun. No matter what we touch, an atom or a cell, we cannot explain it without knowledge of the universe....What am I? What is the task of humanity in this wonderful universe?"[1] Much before, the pre-historic Vedic Aryans upon close observation of cosmic order and natural phenomena had realized that the man within is no alien to the Nature. They discovered the natural laws that uphold the universe, which reflect not only matter and energy but also life, mind and consciousness.

> I know this all embracing Truth
> Resplendent like the sun.
> This is beyond darkness.
> Knowledge of this Truth alone
> Leads man infallibly to immortality.
> This is the only path to be trodden.[2]
> To escape the jaws of death.
> This changeless alone is to be known.
> This is the Truth ultimate.
> This is Immanent in the heart of all.
> There is nothing to be known beyond this.[3]

The Vedas constitute what could be called a 'science' in the modern sense of the word, and much more. In Vedic science, the universe is a manifestation of consciousness. Our true self is not merely the human or psychological self but the universal self. Each one of us is a unique manifestation of the universal consciousness, a human embodiment of it, but our true being is one with the entire universe, with all beings, and ultimately with the absolute beyond all time and space. The Vedas say:

That alone is the Fire, that is the Sun.
That is the Air, that is the Moon.
That alone is resplendent.
That is consciousness all embracing.
That is liquid and the lord of species.[4]
That which is one and without second
Has become this all.[5]
This resplendent truth is shining ad infinitum
Encompassing the Earth from every direction
It has evolved as life on earth.
It is evolving and shall evolve in future.[6]

The Vedic seers worshipped the forces of Nature, consciously realizing that the forces of Nature are the manifestations of the universal consciousness.

"The all embracing truth innermost
The terrible laws of nature
The will indomitable to act scrupulously,
The conscious strivings to the end,
The prayers surging from within
And the sacrifices selfless and benign –
These six alone maintain the Earth."[7]
"Earth is Mother,
And I am the son of Earth"[8]

This is what the Vedic seers had done to ensure their living in conformity with the laws of nature, which is now called "Ecology", derived from the Greek word 'oikos' which is derived from the Sanskrit word 'Okas' meaning shelter/home/plant, defined as the scientific study of the relationships of living organisms with each other and with their environments. The concept of 'one with the nature' is as old as the Vedic age but the relevance is felt acutely now when man searches out totally new ways to anchor himself, for all the old roots-religion, nation, community, family or profession are shaking under the hurricane impact of accelerative thrust.

According to evolutionary history, humans are an integral part of the natural ecosystem. Before the industrial revolution, all organisms-including humans remained an integral part of a complex yet well-synchronized food chain driven by natural forces. An important departure, however, has been the increase in human population beyond the holding capacity of a natural ecosystem pyramid. Agriculture was developed to feed the ever-increasing human population. Agriculture destroyed much of the ecosystems. Humans developed techniques to produce machines and industrial goods to make life and lifestyle comfortable. Industrial civilization depleted Earth's finite natural recourses extensively. The pristine rain forests of the Amazon region in Brazil are now modified in the name of development. Much of the global land, water and air are now polluted and/or contaminated with toxic chemicals released from mindless industries in the mad rush to boost productivity and profitability. Such activities are often against nature or in other words self-destructive. The nature's way of doing things are quite balanced; resources continue to recycle[9] and never deplete and all varieties of life are interdependent with the population of each species under check and balance imposed by nature. Man's attitude of wanting to remain dominant may prove very costly threatening the very existence of man on Earth. As an economics-centered society, the West legitimizes its action on land, water, sky and the biosphere. But, the Eastern cultures do not attribute a higher value to humans than to the remainder of the organisms. To quote Veda:

> Thou art woman,
> Thou art man,
> Thou art boy,
> Thou art maiden
> Thou art old man tottering with the staff.
> Thou exist in all forms.[10]

Human knowledge gained over several millennia, through testing and conjecture and through observation and interference is a priceless legacy of human existence. The material science of the West and the spiritual science of the East should find a common platform of action to pull man from the ecological crisis he has landed himself on. We have to act within a framework where every organism is valuable and that nature and self are but one. Such a realization shall surely provide us necessary scientific methods of solution to address important issues of natural imbalance, civilization diseases, violence, inequality, poverty and hunger.

Let us all pray that wisdom shall prevail upon man and his religion shall teach him that "Let there be peace and balance everywhere".

Dyouh Shantih. Antarikshyam shantih. Prithvi Shantih. Apah Shantih. Oshadhayah Shantih. Vanaspatayah Shantih. Visve devah Shantih. Brahma Shantih. Sarvam Shantih. Shantireva Shantih. Sa ma Shantirdehi. (Sukla Yajur Veda-36-17- Page 584).

Notes

1. Around 2000 B.C.
2. Svetasvataropanishad-3-9, p. 137.
3. Svetasvataropanishad-1-12, p. 135.
4. Sukla Yajur Veda-32-1, p. 530

5. Rg Veda-8-58-2.
6. Sukla Yajur Veda-32-4 p. 531.
7. Atharva Veda-12-1-1.
8. Atharva Veda-12-1-12.
9. C, H, N, P, S.etc.
10. Atharva Veda-10-8-27.

DAWN OF A NEW RELIGION–NATURE

Anjali D'Souza

Introduction

In the beginning was the Word and the Word was with God, and the Word was God. (Bible, Jn1:1). With Him are the keys of the unseen that no one knows but He knows whatever there is on earth and in the sea. Not a leaf falls but with His knowledge: there is not a grain in the earth shadow, not a thing, freshly green or withered, but it is in a clear record (Sura 6:59 Quran).[1] I pervade the universe, all objects in the universe rest on me as pearls on the thread of a garland (Shri Krishna in Bhagvad Gita).[2]

The holy texts explain the beginning of the universe and life by a creator, God, Yahweh, Allah, etc. However, from the scientific view, man, the latest product of organic evolution on earth, is awed by his own origin. There are countless definitions of life, but life slips away from each one of them, because there always seems to be one or more exception. Throughout the study from a simplest atom to complex living organism, it is impossible to say where non-life ends and life begins says-Aristotle.[3] The study of life is a living as well as a loving science. Life itself is dynamic, cannot exist alone but is entwined with the whole creation. Life needs an environment to live in. There is an active equilibrium between the living organisms and their surrounding. This equilibrium is fragile, and if destroyed can destroy life itself.

However, man, the supreme of all living beings has forged ahead and has destroyed it in more than one ways shaking the existence of life around him and himself too. In human experiences with nature we see both are best and worst capabilities. The best includes *why despite warning science forms a stressed global ecosystem, mounting scientific evidence and public education campaigns, why does degradation of the environment continue to persist and mount?* But the worst capability is destroying the fragile equilibrium for our own use. Securing food supplies for the next generation depends on an all out effort to stabilize our population and climate but we resist change in our reproduction behavior and shift from fossil fear based economy to solar/hydrogen based one.

The holy texts have instructed humans to keep in mind the fragile equilibrium which exists around him sustaining life. Hinduism is a religion that is very near to nature. It asks its followers to see god in every object in the universe, worship of god in air, water, fire, sun, moon, stars and earth is especially recommended. Earth is worshipped as the spouse of god, hence very near and dear to God. All lives are considered as children of god and earth.[4] The earth is mentioned some 453 times in the Quran whereas sky and the heavens are mentioned only about 320 times. Islam does understand the earth to be subservient to humankind but it should not be administered and exploited irresponsibly. "Do you not observe that God sends down rain from the sky, so that in the morning the earth becomes green?"(Sura 22:63). The color green is the most blessed of all colors for Muslims and, together with a profound sense of the value of nature as God's perfect and most fruitful plan, provides a charter for a green movement that could become the greatest exertion yet known in Islamic history a "green jihad" appropriate for addressing the global environmental crisis.[5] All of the earth community is valuable to God, who continues to create, sustain, and redeem the whole.

God relates directly to and cares for the well being of other kind, created to enjoy being in their own right and not only function as companions or helpers of humankind. *Christians are Gods spirit in creation, God is the power of life giving breath, the wisdom continually working to transform and renew all life and the love that sustains it.* Biblical images portray the Spirit as "a healing and subversive life-form as water, light, dove, mother, fire, breath,...wind" and comforter of suffering.[6] Christian theology engages in "psychology" because it is study of the psyche or soul as much as it is study of God and God's revelation. However, despite religions advocating the care of nature humans have forged ahead and are in the process of destroying the balance of nature and in the process staking all life in the universe.

Present Scenario

In the present industrial and technological age, as most people already know, humans are not living sustainably. Relations to the environment and consumption of resources are significantly different in developed and developing countries of the world. People do not bear equal responsibility for ecological abuse, nor do all relate to nature in the same way. The question requires exploration of the complex factors shaping how people represent, value and interact with their environment. There is life that is inferior. All lives enjoy the same importance in the universe and all play their fixed roles. Humans are now trying to satisfy that need through violence, obsessive work, indulgence in sex or food and use of drugs.[7] They are to function together and no link in chain is to be lost. If some link is lost, the whole ecological balance would be disturbed. All kind of life- birds, insects, and animals- contribute toward the maintenance of ecological balance, but what is humanity's contribution towards it? Humans are intelligent animals; therefore, our contribution should be the biggest. We are nullifying the benefits of the contributions made by other species of life because of our greed for material enjoyment and our craze for power. We do not allow earthworms and bacteria to maintain fertility of the soil by using chemicals, manures and insecticides that kill them. We destroy plants and forest indiscriminately and come in the way of plants providing oxygen essential for the very existence of life. By destroying plants and forest we become agents for increasing the deadly carbon dioxide in atmosphere. We pollute the air by burning oil in all kinds of machines. We produce unhealthy sounds through various machineries and instruments which cause sound pollution. By building towns and cities on the banks of rivers we pollute the rivers.

The present environmental crisis is well documented in its various interlocking manifestations of industrial pollution, resource depletion, and population explosion. The problems are manifold, namely, the essential ingredients for human survival, especially water supplies and agricultural land, are being threatened across the planet by population and consumption pressures. With the collapse of fishing industries and with increasing soil erosion and farm loss, serious questions are being raised about the ability of the human community to feed its own offspring. Moreover, the widespread destruction of species and the unrelenting loss of habitat continue to accelerate. The environment crisis presents a serious challenge to the world's religions. This is especially true because many of these religions have traditionally been concerned with the paths of personal salvation which frequently emphasize other worldly goals and reject this world as corrupting.

Need of the Hour

This is the 21st century which is marked by both major progress and major problems. In such situations awareness is very important. It does not matter whether we believe in God or in a next life; we need to create a balance between science and spirituality. If the two remain distant we head for trouble. In a riverbed were two trees. One was a mighty banyan tree and the other was a reed. "You are spineless; look at me not only am I tall and stand erect, I also give shade to the weary", it would often say to the reed. In the rainy season came a flash flood that uprooted the banyan tree. The reed however survived because it simply bent with the current and when the floods receded it became erect again. This is a story of the strength of humility. Another great lesson the story contains is that it is only those systems that come in equilibrium with the surroundings that survive.[8]

Coming into equilibrium with the surroundings also means actively interacting with the surrounding forces like sun, wind, atmosphere and gravity. A system can only interact with the surrounding forces when it can sense them. Thus, natural systems have developed mechanisms for sensing all these elements and hence, have temperature, humidity, solar, chemical and gravity sensors. It is necessary, therefore, to stimulate and sustain the *"ecological conversion"*, which over these last decades has made humanity more sensitive in facing the catastrophe toward which it was moving. Man is no longer "minister" of the Creator. However, as an autonomous despot, he understands that he must finally stop before the abyss. "Another welcome sign is the growing attention being paid to the 'quality of life' and to 'ecology', especially in more developed societies, where people's expectations are no longer concentrated so much on problems of survival as on the search for an overall improvement of living conditions". Therefore, not only is a "physical" ecology is at stake, attentive to safeguarding the habitat of different living beings, but also a "human" ecology that will render the life of creatures with more dignity, protecting the radical good of life in all its manifestations and preparing an environment for future generations that is closer to the plan of the Creator. How to adapt religious teachings to this task of revaluing nature so as to prevent its destruction marks a significant new phase in religions.

Thomas Barry, a historian of religions has pointed out that what is necessary is a comprehensive revaluation of human earth relations. The new relation which is spoken of is ecopsychology. It is based on the presupposition that people are bonded to nature as they are bonded to their families. It is seen as a positive motivator by helping us recognize that we just don't need the earth to survive–like some planetary service station we stop at long enough to suck up fuel–we also love it. Ecopsychologists believe that we have an innate love for the environment, we need to uncover it. If this science is studied there are reasons for environmental degradation and methods to restore the planet to what it was as before. The ecological circle consists of a healthy relationship between humans and earth. Nature also refers to "human nature". "Mother nature" and "human nurture" are bridges that focus on human relations with environment. In Ecopsychology nature is defined in two senses- the human nature and more than human nature that concerns ecological scientist, thinkers and activists. Any Fisher, defined Ecopsychology as a subject that is concerned with bringing together of these two natures and study their interrelation.

Dawn of a New Religion

Environmental crisis, cultural and political processes have created a deep and lasting impact on the human psyche and soul. The inner and outer worlds mirror support one another and hence a healthy ecosystem is inseparable from healthy psyche. According to Pullitzer prize winning sociobiologist E.O. Wilson people have an inherent need to be in contact with the out-of-doors which is known as 'Biophilia'. Nature holds the key to our aesthetic, intellectual, cognitive and even spiritual satisfaction. Through nature therapy, or ecotherapy this biophilia can be harnessed and people can be connected to nature. Once connected people strengthen their conviction towards environmental issues, take more seriously their responsibility to live lightly on the earth and strengthen their health.

Daniel C. Noel a psychologist of religion in his paper focuses on the human psyche and the human natural world. The most basic ingredient to restore the world is when we think in a full-fledged NATURE RELIGION.[9] Degradation has been caused due to alienation from nature. The human conceptions about the world, mankind and their interrelationship are the mental and psychological causes of the ecological crisis of our times. These ultimate causes ought to be described as macro-ecological. The macro-ecological factor i.e. the human mentality is part of the human being; it is the God-created personality, with its special and precious abilities and charismata, incorporated in it "after God's image". So, the state of the mind (mentality) and the condition of the human being are inseparable. Therefore, when we speak of the human mentality, we mean the spiritual condition itself. From this point of view, it is clear that we see the function of the human mind as an intellectual activity strongly and inseparably related to the Creator and to His creation. Strictly speaking, the ecological crisis is not a question of mankind's relationship to nature alone, but a question of relationship to the Creator and His purpose in creating the world. This is the macro- ecological view of the problem. The micro-ecological aspect is man's activities and use of the world, as side-effects of the macro-ecological human view of the same. Today the ecological crisis has assumed such proportions as to be THE RESPONSIBILITY OF EVERYONE. As it has been pointed out, its various aspects demonstrate the need for concerted efforts aimed at establishing the duties and obligations that belong to individuals, peoples, States and international community. This not only goes hand in hand with efforts to build true peace, but also confirms and reinforces those efforts in a concrete way. When the ecological crisis is set within the broader context of THE SEARCH FOR PEACE within society, we can understand better the importance of giving attention to what the earth and its atmosphere are telling us: namely, that there is an order in the universe which must be respected, and that the human person, endowed with the capability of choosing freely, has a grave responsibility to preserve this order for the well-being of future generations. THE ECOLOGICAL CRISIS IS A MORAL ISSUE. Ecopsychology combines the human element from psychology with the study from biological systems. It helps humans to experience themselves as an integral part of nature.

Means and Actions

Humans are relational beings. Our identity is shaped through experiences with others. What does it mean to have an identity as an individual, yet always to exist in network of relations? There are things about ourselves that we do not easily know. There is an element of depth in human psyche. There are two tools of analysis (1) Religious analysis (2) Psychological analysis. When our new religion is nature it will be derived from a bond of love and loyalty to nature. Keeping this in mind each human will be responsible to others and also to the environment. The sustenance of ecology can be done by energizing spiritual awareness that motivates and empowers us to engage in opening ourself and participating with others actively to help save the environment. The goal of ecopsychology is to use nature to promote healing and growth; healing by wilderness treatment by smelling the earth, seeing the sunlight and hearing the babbling brooks such therapies are involved in healings termed as ecotherapies. Other types of ecotherapies are horticulture therapies, use of trees and animals. There is certainly something very powerful about wilderness, it evokes strong emotions, responses, and fantasies. Theorist gains a more complex understanding of representations of nature and motives for environmental abuse by exploring the psychological and theological territory of the "wilderness within human soul". But this individual territory is always also the social. It is both the psychological and the social dimensions of environment issues and the interplay between these dimensions.

Education can be one of the major areas of thrust, persons involved through ecopsychology learn to bond with their natural world and lead an eco-friendly life and then educate others. The primary way of saving the ecology is by having a constructive ecological life style. Ecoeducation therefore will teach earth literacy, earth caring and will be rooted in earth bonding. This will cause persons to save the environment because of their deep rooted love and loyalty to nature. This is also called the *self earth care plan.*

Conclusion

Environment can be saved if certain key issues are addressed to save the earth such as population growth, sustainable economic growth, being nurtured by nature, humans changing their culture to become more earth friendly appreciating the gifts of earth valuing the cycles of nature.[10] As F. Bender says in a review of Ecophilosopher J Baird Callicott's *earth's* insight "Its chief shortcoming, in the reviewer's opinion is one that is endemic to environmental ethics per se, a discipline that seems to wish only that everyone would stop harming the Earth in order that all our ecological problems might be solved". This new approach of environmentalist can help save the world and humans. They combine ecology, psychology, and religion to form a new religion: nature for helping the world wisdom for our age. Therefore, nature can be addressed as a "Good Mother"- a beneficiate and nurturing mother who provides and maintains all life. Mother usually evokes traditional meaning of a bountiful, self sacrificing power who herself requires no care. But it can be a "Bad mother" too. Nature can become the violent, threatening, wrathful mother, when humans and their technology master unruly version of Mother Nature. Today "Nature as Hurt Mother", looks at possibilities and pitfalls of environmental healing in the image of nature as the mother we have wounded and now seek to heal.

Notes and References

1. Frederick M.: Islam and Ecology: A bestowed trust inviting Balanced Stewardship, Introduction to Islam, Denny, university of Colorado.
2. Rao Sheshagiri: Hindu Faith Statement: Encyclopedia and Hinduism.
3. www.encyclopedia.mu/encyclopedia/e/ec/ecology.html.
4. Rao, Sheshagiri: Op.Cit.
5. Kaveh L. Afrasiabi: Towards an Islamic Ecotheology Hamdard Islamicus, Vol. XVIII, No. 1, 1995, p. 40.
6. Dieter T. Hessel: Christianty and Ecology; Wholeness, Respect Justice, Sustainibility, Programme on Ecology, Justice and Faith..
7. Palo Alto C.A: Foundation for Global Community. Bayland productions Producer, 1995. "Ecopschology, Restoring the Earth, Healing the Self" (Videotape), Pato, CA: Foundation for Global Community.
8. www.indiatimes .com, Speaking Tree.
9. Daniel C. Noel Nature: "The Necessity of an Ecopsychology of Religion, Nature Religion as a Theoretical Construct Reflections from an Emerging Field." held at American Academy of Religion Annual Meeting, Orlando, Florida, 23rd November, 1998.
10. Dash, M.C., Fundamentals of Ecology, Tata McGraud Hills Publishing Co Limited Pp. 1-2.

SACRED PLANTS TO AROUSE SPIRITUAL CONSCIOUSNESS AND SUSTAIN ECO-BALANCE

G. Therasamma

'God's holy life would not be manifested without Nature' (Jacob Boehme)

In India, and perhaps in many other countries of the world, trees have been held in veneration since ancient times. Our Rishis of yore retired to forests to meditate and to do penance amidst trees and "far from the maddening crowd'. Our forebears loved trees and forests so much that they even worshipped them as abodes of Vrikshadevata (The Goddess of Trees) and Vanadevata (The Goddess of Forests). Many religions and cultures have been rooted in forests and elsewhere, not through fear and ignorance but through ecological insight. A natural system of conserving is in vogue in Himalayas, almost every hilltop is dedicated to some local deity and the trees on or about the spot are regarded with great respect, so that no body dare touch them.

The original concept of tree worship however, finds mention in Vedic literature. It was promoted in the Mahabharata, saying, "Trees are to be worshipped". In Catholic religion plants are used symbolically. In Jain and Buddhist religions worship of trees, was of great significance. In eastern rural areas trees and plants have become an integral part of social life. Most of the religious festivals include worship of trees. Faith is the root cause of such tree worship, on realizing the goodness of tree and plant. So they are honored as holy. The trees that we worship are known as sacred trees. These sacred trees are seen as an image of the cosmos, a symbol of the inexhaustible source of cosmic fertility.

Views of Different Religions on Nature

The relationship between mankind, the universe and God or the Universal spirit is the foundation of *Hinduism*. Ecological sensitivity in Hindu tradition made people to extol the five fundamental elements–the earth, the water, the fire, the air, and the space as the powers of the natural world. The prayer of the great Rishis of India 'Loka Samastha Sukhino Bhavantu' explains the importance of every one and every thing in the universe'. Vedic Gods were embodiments of nature and its forces such as Dyaus, Prithvi and Varuna. Hindu culture depicts the tree as a symbol of abundance and veneration. Besides trees and rivers, rocks and mountains, indeed all forms of nature were believed to have spirits to be respected. All the traditions and customs of worshipping nature in Hinduism depict an ecological insight. It is a natural system of conserving and instaining the environment for ecological balance. The festivals Vijayadasami, and Sankranthi are the practices of preservation and transmission of culture in ecological perspective. Practices and traditions of Hinduism expanded to formulate an ecological consciousness.

The very foundation for environmental protection in the *Islamic* faith is the religious principle that Nature is the creation of God and human beings are obliged to obey Him. Prophet Muhammad said, "When doomsday comes, if someone has a plant in his hand he should plant it as it is to serve God for mercy and generosity. If life on earth is preparation for eternal life in heaven, then the loving care of the natural environment would seem to be appropriate training. Muslims have a profound sense of the value of nature and the colour green, the most Blessed of all colours provides a charter for a green movement "Jihad,' which is appropriate for addressing the global environmental crisis.

Buddhism is against the violent attitude of the human beings towards nature. A close relationship between human morality and natural environment has been systematized in the theory of the five natural laws. Siddhartha was born in a grove of Sal trees and attained enlightenment under the peepal or Bodhi Vriksha, tree of 'wisdom and knowledge'. A good number of texts demonstrate ecological insight and no single text encompasses the diversity of Buddhist doctrines and practices related to the environment. The teachings of Buddha reflect the ways of life with environmental consciousness, which inspires and guides the present day environmental issues. The classic sources are often re-interpreted and thus Eco-Buddhism continued to develop. A clean environment enables the development of proper moral and spiritual life style. There is a reciprocal relationship between man and nature. There is a direct effect of man's nature on the physical nature, like the greed, hatred and elusion in man produce pollution, and generosity, compassion, and wisdom produce purity.

The bio-cosmology of *Jainism* presents a worldview that stresses the inter-relatedness of life forms. To Jains every thing has a soul. Reality consists of living and non-living substance. Contact between two results in Karmic entrapment. *Japanese* are likely to believe that one should live in harmony with nature and recognize the ecological devastation. Nature is idealized in its "cultured" forms. *Shinto* beliefs and attitudes towards nature are relevant to the problem of environmental preservation and include sacred space and time, relation between nature and human beings and idea of purification to re-establish order and balance between nature, humans and deities.

Rich diversity of perspectives about God's relationship to nature is presented in the *Judeo-Christian* tradition. God blesses creation and it praises God. One must honour God's creation and value it, because nature, the creation of God is precious to him. A Jewish tradition believes in the kinship of all creation and shared mission of serving God, each creature in its own way is often compared to a cosmic song. In Jewish tradition people used to pray under the 'fig tree.' Date palm according to Jewish tradition is their 'Tree of Life.' Apple or Apricot tree is regarded by Christians as their 'Tree of life and knowledge'. The 'fig-tree', which bears no fruit, is a symbol of unbelieving Israel. The Olive-tree branch has been regarded as the symbol of peace. Jesus says, 'I am the Vine and you are the branches', as a symbolic of deep relationship. The prophets of the Old Testament and Jesus in the New Testament used trees, vines, lilies of the field, thistles, bitter herbs, grapes, hyssop and dozens more plants to symbolize God's mercy, bounty and judgment. Their products were used for physical healing in biblical times and can be used today for physical and spiritual healing.

Taoism says that everything in nature is created on equal footing and man receives no special status. Man and nature should go hand in hand and it teaches us the inter-dependence of all life forms and non-living things. The 'World Tree of the ancient Chinese was that of Peach'.

The tribal, *Indigenous Community* perspective recognizes the importance of our life, including the life of all organisms on the earth. The earth according to the Tribals, world over, is a 'living conscious being' that must be treated with respect and loving care. The Bishnoi community (a community famous for its passionate dedication for regional flora and fauna) of Rajasthan is ecologically aware people who for generations have been practicing environmental conservation and wise resource management.

This is the kind of outlook we had been having characterized by genuine spirituality which formed the essence of wise ecological consciousness. In spiritual dimension, when we act as self-absorbed individuals, with little regard for anyone or anything that exists outside ourselves, we immediately fall into moral and spiritual error. Therefore, there should be a facet of ecological awareness that should not be overlooked.

Relationship of Trees with Devtas

Vegetation is not only the gift of nature but it also unites us to the Almighty. In ancient times the sages lived in forests for studies, meditation and penance. Mankind developed and advanced its process of learning and knowledge in forests that promoted feelings of kinship with plants and trees. Vegetation has been considered to be an equal partner in one's joys and sorrows. The practice of Vanabhojana to pray to God for the rains is symbol of belief and reverence to trees. The advent of spring, with greenery due to rains, is a period of festivity with intimate feelings, to Indians even today.

The ancient literature has established the greatest fact that life, which pulsates in us, also prevails in plants and trees. Therefore, plants are not separate beings but they are just like us. Garhwali folk song says that 'one's own son is as important as other man's tree' hence, both should be guarded from any damage. Food is the be all and end all of all the creatures. All beings that exist on earth are born of food. 'Taitareya' All the living beings emerge from grain. No activity is ever possible without food. That is why God says 'He is food', Annam Parabrahmam'. Therefore food, good universal medicine, in right dose blesses, and in over dose kills. The vividness and vastness of vegetation reveals the splendour of nature.

Flowers are the most valuable gift of nature to mankind. The most important fact is man's psychological feeling towards flowers. Nature has not bestowed such a gift of faith, aesthetic sense and happiness to other beings. Many flowers are used for purification; they please the mind and have the capacity to purify one's soul. Scientists have accepted the truth that fragrance of flower has heavenly and pleasant effect on the brain cells. The mind gets unimaginable peace. It is accepted that fragrance of a flower helps in getting sound sleep. It is said that it releases high blood pressure and helps to get enough relief.

Trees that are Worshipped in Indian Tradition

Banyan tree is believed to know the secret of the heart, and no one can dare speak a lie under it. In this way the vegetation not only gives us joy and pleasures but also tends us philosophically and helps in uplifting our soul and developing spiritual consciousness.

- KHEJRI: *Prosopis spicigera* (Jammi-chettu): It is worshipped on "Vijayadashmi" day in almost all parts of Bharat. It holds the bow of Arjuna and calls Ram as dear one. It destroys whatever is inauspicious, evil and fulfills all desires. The wood of Jammi has proved to be the best for sacrificial ceremony (Yagya). It is also worshipped as a symbol of fertility.
- PEEPAL: *Ficus religiosa* (Bodhi tree): The peepal tree is considered very sacred and worshipped by the men and women folk for the fulfillment of their desires. It has the power of immortality hence it is named as Mahadev and it is believed that it bestows long life and health to those who worship it. The special significance of the tree is that it helps by releasing oxygen and purifying the atmosphere.

- BILVA: *Aegle mamelos* (Maredu tree): Bilva means 'that which destroys disease'. The leaves of this tree are used for the worship of Lord Shiva. By offering Bilva leaves to Lord Shiva, one is bestowed with immense wealth, and divine majesty, grandeur, prosperity and affluence.
- HOLY BASIL: *Ocimum sanctum* (Tulsi): Tulsi occupies the most respected and sanctified position from the point of view of health, religion, metaphysics and even has decoration value. Tulsi is indispensable in any religious ritual or act of worship. Tulsi used to be worshipped in the rituals of the Greek (Eastern) Church and every year on the Birth Anniversary of Saint Basil, women used to sanctify their houses by scattering Tulsi leaves which had been previously offered in churches, inviting the Saint's blessings to bestow good fortune on the family in the new year.
- MONKEY BREAD TREE: *Adansonia digitata* (Kalpa Vriksha): In ancient times the concept of 'Kalpa-Vriksha' was one of the main amongst other pious, aesthetic and heavenly trees. It is a tree that grants all desires. Mahabharata, Ramayana, Jatak grantha (Scripture) and Jain literature give vivid description of this tree.
- MANGO: *Mangifera indica*: There are references in ancient Sanskrit literature to the Mango as the abode of the koel (Indian Nightingale) and to Kamadeva, the Love God. Twigs and leaves of the tree are used on festive occasions as a sign of auspiciousness.
- BANANA: Banana references occur to this tree in the epics of the Pali Buddhist Canon of 500-600B.C. On auspicious occasions, especially in the South, the cut banana trees adorn doorways.
- SANDALWOOD (Chandana): Sandalwood is among the elite and most renowned of sacred trees. The famous Murugan temple at Tiruttani, Tamil Nadu, has a rare stick of sandalwood. It is ground for paste used in the temple rituals, which is then given to devotees. This special paste is lauded to have curative powers. But the mystery of this temple cure is that the stick of sandalwood is never to shrink.
- *Cannabis ganga* The Sacred Rivers of the Sadhu: Cannabis has been a five-purpose plant fibre, seed oil, for its seed as food, for its psychoactive properties and therapeutically as a medicine. Tradition in India maintains that the Gods sent man to the Hemp plant so that he might attain delight, courage, and have heightened sexual desires.
- *Papaver somniferum* The Poppy of the Holy Mother: The Opium Poppy, which is one of the most medically important plants known to man which still plays a central role in the control of pain and suffering particularly in terminal conditions shows a very early pattern of use and cultivation and scalp complaints. Opium was taken by participants in certain religious ceremonies during the late period to induce a state of ecstasy essential for the performance of the sacred rites. Egyptian

Medical texts list among Opium's many uses its sedative powers to alleviated the pain of wounds, abscesses.

- DATURA: Since time immemorial various Datura species have been revered as sacred visionary plants practically by all cultures that have come into contact with it. They have power to induce visionary dreams, to see the future and reveal the causes of disease and misfortunes. *Datura ceratocanla* is used in discovering the whereabouts of lost or stolen objects, divining the outcome of future events and prognosticating the causes of diseases, especially if black magic is suspected. *Datura fastuosa* has the clairovoyant use. The ointment prepared out of its species is used as an ointment for cracked soles, injured feet, infected wounds, skin sores and to treat ulcers and rheumatic aches and pains on one hand. All species of datura have also the power to kill in sufficient quantities.
- THE SACRED MUSHROOM: Teonan and the lady of the Alder: All plants are regarded as having spirits or souls. Some plants are personified as deities, which must be treated with care and propitiated with offerings. These are called entheogens that means "the God Within.
- RUDRAKSH: It is believed to have been born from the tear of Lord Shiva. The fruit of this tree has high religious significance and wearing this on one's body results in a number of advantages like 'sound mental health and physical health. Paul Hume, who conducted research for 20 years on Rudraksh fruit, went on and said that wearing this could make the impossible possible.
- Beetle leaves, which are used in many auspicious occasions, have also got a lot of health and medicinal value. These leaves provide Vitamin 'A' to the body. Beetle leaves reduce plum in the lungs, removes bad odour in the mouth, increased brightness in one's body.
- Mango leaves, apart from being signs of festivity these are used in auspicious ceremonies and they are aesthetic to the onlookers and they even ward off pollution in the atmosphere. These mango leaves, along with neem leaves, are tied in a garland like manner and tied on the thresh holds of the houses.
- Tree worship continues to this day among the devout and the religious, and among our rural folk. Pride of place goes to the Pipal, which is revered in almost every part of India. Sometimes the Pipal and the Neem are grown together and the two are ceremoniously married.

In tribal tradition sacred plants only provide the key to other dimensions but what is to be found there depends on the mental state of the initiate on the psychological, emotional and physical preparation and purity. Besides these there are so many other trees like–*Azadirachta indica*, Oak plant, *Bauhinia variegata*, Banyan, *Citrus auranntifolia*, *Emblica officinalis* gaerln, *Butea monosperma*, *Cocos nucifera*, Asoka, Kadamba, Vakula, and Jand, which are worshipped or treated holy because of their medical and other values.

Sacred Groves

Sacred groves, dedicated to local deities or ancestral spirits, are found all over the sub continent. Local communities protect them from over-use. Thousands of sacred groves in almost all the states have been documented, ranging from a few trees to forests of many acres. Such sacred groves with the almost 40,000 endogamous groups are of increasing interest to native conservation because they help preserve threatened plants and animals. Environmentalists are working with local communities, recognizing that traditional knowledge and sacred practice are important elements in the conservation of environment. The groves, besides harbouring plants and animals, are crucial for soil and water conservation. Due to urbanization and pressures of population, many sacred plants are rapidly becoming endangered. Our generation must be committed to preserve these traditions of conservation for prosperity.

Reported Number of Sacred Groves in India

State	*Number of sacred groves**	*State*	*Number of sacred groves**	*State*	*Number of sacred groves**
Andhra Pradesh	750	Jharkhand	21	Rajasthan	9
Arunachal Pradesh	58	Karnataka	1424	Sikkim	56
Assam	40	Kerala	2000	Tamil Nadu	448
Chhattisgarh	600	Maharashtra	1600	Uttaranchal	1
Gujarat	29	Manipur	365	West Bengal	670
Haryana	248	Meghalaya	79		
H. Pradesh	5000	Orissa	322		
Total					13,720

Sacred Plants Resort developed near Bangalore include Narasimha *vana*, Sankashta Chaturti *vana*, Vara Lakshmi Vrata *vana*, Ashoka *vana*, Navagraha *vana*, Kanva *vana* , and 25 such designated areas. Each spot is designed in a way to represent, enhance or reflect the nature; religious back ground and sacred aspect of each plant. This Sacred plants resort is expected to create an atmosphere to become one of the strongest weapons of afforestation to motivate people to develop respect towards forests and plants.

Sustained Eco Balance

Awareness of the loss of natural harmony brought an awareness of fostering sustainable human-earth relations, sensitivities to local regions and their bio-diversity for future generations. Indigenous environmental knowledge is the felt-experience of interacting with the larger whole of reality and an understanding that spiritual forces capable of leading humans into both utilitarian and self-understanding abide in the places.

The crying need of the hour is that tree-mindedness must get into the roots of the human psyche. Trees are bound by mutual relationship with water, creatures and animals, soil and atmosphere, fuel, industry, food and fodder, fertilizers and medicinal attributes. In the primitive stages when human beings lived mostly in jungles, trees provided shelter and food to them and thus carved out a way to the present day civilization. By destroying this wonder gift of nature we are only disturbing the balance on the earth resulting in damaging consequences. It takes 30 to 40 years to produce full-grown trees, and little do we realize how much of ecological imbalance we would be causing by cutting down a single tree. Thus in a way by cutting the trees we are destroying ourselves. Trees remain man's friends throughout their life, so let us all try to be their friends too.

In an incident in America, a scientist working in his research laboratory accidentally cut his finger with a blade and a plant kept in the lab started showing its feeling of grief and sorrow. The mental agony of the plant was recorded on a graph. The affectionate behaviour of the plants, their description of revealing the sentiments of man is just a glimpse of their interesting world. The fact is that the whole world of plants and trees is full of human feelings. The need of the day is to recognize and understand these feelings.

Trees being worshipped by people are helping man in maintaining the eco-balance. They play a significant role in purifying the environment by taking in carbon dioxide and releasing oxygen. Scientists perceive that one hectare of green vegetation provides us approximately 600 kg of Oxygen in barely 18 hours and absorb 900kg of Carbondioxide. A tree weighing 50 tons gives about 1 ton of Oxygen, a year. They effectively control 25% of pollution and poisonous gases. This attribute is specially found in Peepal, Neem, and Tulsi plants. Plants also play an important part in the cycle of nature maintaining its balance. Thus worshipping trees helps a lot in having a good environment all around the man.

In every tree, plant vegetation enshrines a unique power of nature and a mystery written in its folds the knowledge and understanding of which enables the individual to be capable of unfolding the mysterious layers of unitary and plurality. Vanamahotsav Movements are vigorously observed in almost every state of India to increase plant life. Tree plantation as a regular programme by the educational institutions is a valuable contribution for conservation of plants. Awareness in the public should be created about the benefits of forests and tree plantation.

Conclusion

Man's greatest concern today is the loss of traditional wisdom and practices brought about by the increasing presence of westernized urban culture and ever expanding market economy. In the modern Educational system, younger generations are not learning to respect the trees and their many values. It is high time we, educationists realize that development, and progress and modernity do not mean turning one's back on the sacred traditions. The need is to integrate this wisdom into our educational systems. The whole universe is one consciousness.

"Issa Vaasyam Idam Sarvam". (Isha Upanishads)
'This whole World is the Abode or House of God'.

References

Agnihotri, S.P. (1995), Environmental Conservation for Global Development, Chugh Publications, Allahabad, India.

Anthony, P. Castle (2001), More Quotes and Anecelotes, St. Pauls, Bandra, Mumbai.

Archana Dange (1999), Garbage to Gardens, Centre for Environmental Education, Bangalore.

Babu J.V. (Ed.) (2005), Know About Plants & Animals–Gyan Vikas Prachuranalu, Vijayawada–3.

Bhoj Raj Dwivedi. Environmental Vastu–Diamond Pocket Books (P) Ltd., New Delhi–20.

Bio Diversity (1998), C.P.R. Environmental Education Centre, The C.P. Ramaswami Aiyar Foundation, Madras.

Chakravarti, S. Venkatesh (1976), "Our Tree Neighbours", National Council of Educational Research and Training, Delhi.

Chaturvedi A.K. (G-228), "101 Essays on topics of Today"–Goodwill Publishing House, New Delhi, India.

<http:environment. Harvard.edu/Religion.html>

<http:www.American Catholic.Org>

<http:www.erowid.org/plants/mushrooms/mushroom-journal.html>

<http:www.odu.edn>

<http:www.Plants cultures.org.uk/themes/sacred and spiritual.html>

<http:www.Plants cultures.org.uk/themes/themes-lan.ding.html>

Isha Upanishads (1500–600 BC) as cited in Bio Diversity, C.P.R. Environmental Education Centre, Chennai. P. 39.

J.C. Kurian (1995), Plants that heal, Oriental Watchman Publishing House, Pune, India.

Joy K.P. (1994), Eco Development and Nature Conservation, AIACHE, New Delhi.

Kelu P. (2000), Environmental Education–A Conceptual Analysis, University of Calicut Press, New Delhi.

Kondas S. (1975), Man made Forests, National Council of Educational Research and Training, Delhi.

M. Madhavi (2006)–Tulsi, Sri Balaji Publications, Vijayawada–2.

Narayana Reddy P., "Aacharalu Sastriyata"–TTD Publications, Tirupati.

Odum H.T. (1990), Environment, Power and Society, John Wiley and Sons, New York, USA.

Fr. Ignacimuthu, S.J. (1998), Environmental Awareness & Protection, Phoenix Publishing House Pvt. Ltd., New Delhi.

Sharma R.A. (2001), Environmental Education, Surya Publications, Meerut, India.

Sita Lakshmi R., The Glory of BHAA-RATH", Sai Shriram Printers, Chennai, India.

Swami Amalovananda , "Atma purna Anubhava"–Anjati Ashram, Mysore, India.

Vandana Shiva (1988), Staying Alive: Women, Ecology and Development in India, Kali for Women, New Delhi.

HONORING LIFE-THE PROSPECTIVE OF FAITH AND SCIENCE

Jayasree K.

> *"The Earth teaches us Patience and Love; the Air teaches us Mobility and Liberty; the Fire teaches us Warmth and Courage; the Sky teaches us Liveliness, Broad-mindedness, the Water teaches us Cleanliness and Purity".*
>
> **– Holy Sikh Book, The Guru Granth Sahib**

We can see Ecology as spirituality, and again spirituality as ecology. Spirituality, a wider concept than religion, embraces ideologies, attitudes and actions that motivate humans in their quest for deeper meaning and experience about life. Concern for the ecology, for the welfare of the planet, may be functioning at the present time as an avenue for development of global spirituality. The ecological problems are real everywhere and also growing at a high altitude.

It's now increasingly becoming evident that it's the human mind that holds the key to the survival and flourishing of the biosphere - noosphere or its possible destruction. It is in this context that the ethical positions adopted by the humans with respect to the other members of the biosphere and to the natural systems as a whole, assumes great significance. Hence the path from ecology to spirituality is a path of enlightenment, a path towards understanding the divine destiny of human kind to protect and to take care of the earthly paradise. A neglect of this destiny will turn the earthly paradise into an earthly hell.

A famous English scientist in the field of theoretical physics of elementary particles, Fritj of copra understood natures' wisdom as "divine, eternal unanimity of nature and reason, fundamental unanimity of all nature's phenomena" - a kind of world outlook (salvation), characterized by genuine spirituality, which forms the essence of wise ecological consciousness.

Ecological catastrophes, impending or already present, are beginning to awaken governments, private groups and individuals to issues like global warming and nuclear pollution, institutional religion, however is still lagging in bringing its influence to bear on these problems. From this perspective, God has always been "incarnate" in various forms in this world. Nature is always engaged in recycling matter. Nature endlessly composes, decomposes and then recomposes forms of matter both organic and inorganic. Cultures of different religions stamp their own unique personality of following different practices. Life eludes the roots of recycling. With scientific speculation on time becoming more and more complex in a universe with 'no edge of space time' as Stephen Hawking says, the old image of the cycle reemerges in the human consciousness in different ways.

There are common spiritual values and common spiritual insights in all the spiritual practices. Spiritual practices give ecological knowledge that views humans as part of the natural environment, not simply as observers or controllers but as "participants" within the natural environment.

India has, through trackless centuries, provided an ample expanse of friendly pace for an open and ongoing discourse of ideas established the principles of ecological harmony centuries ago-not because the world was perceived as heading for an imminent environmental disaster or destruction, nor because of any immediate utilitarian exigency, but through its quest for Spiritual and physical symbiosis, synthesized in a system of ethical awareness and normal responsibility.

The Vedic Hymn to the Earth, the Prithvi sukta in Atharvaveda, is unquestionably the oldest and the most evocative environmental invocation. In it, the vedic seer solemnly declares the enduring filial allegiance of human king to mother earth. The vedic seers regarded the Earth as 'Sacred Space' for the consecrated endeavors and aspirations of humankind and for the Practice of restraint and responsibility. Such a world can only be sustained by 'Satyam Brhat Rtam Ugram', The severely exacting discipline of truth, harmony and rectitude, based on a conception of cosmic and comprehensive peace as envisioned in the famous vedic Rymn of peace, we invoke and imbibe Aum, the primordial sound of cosmic Harmony and pray for: Peace and Harmony in Heaven; sky, on the Earth; in the waters; the Herbs, the vegetation and the Forests; among the Peoples and the Rulers of the world; in Spiritual Quest and Realization; for one and all; Peace and Harmony in Every thing and Every where.

Some of the Spiritual Practices Followed in Various Religions

Hinduism

Hinduism is one of the most ancient religions of South Asia, dating back at least 4000 years with more than one origin. Scholars of the vedas have held for the various texts and rituals that extol the earth (bhu), the atmosphere (bhuvah) and the sky (sva) as well as the goddess associated with water (Aap), with fire and heat (Agni), and wind (Vayu). They have noted that the centrality of these gods and goddesses suggests an underlying ecological sensitivity within the Hindu practices. In later Indian thoughts, these vedic concepts become formalized into the samkhya denotation of five great elements (mahabhuta); earth (prithvi); water (jal); fire (tejas); air (vayu), and space (akasa). The meditative and ritual processes of Hinduism entail awareness of these constituents of materiality and daily worship (puja) employs and evokes these five powers.

India has a long history of forest protection from the edicts of Asoka, to the individual work of various rajas, to the modern Chipko movement. In ancient times sages lived in forests for studies, meditations and penance and the ancient sages found glimpses of great Rudra in plants and vegetation and considered them as Gods. In order to explain the significance and importance of trees to people as bestows of inspiration for life, pleasure and prosperity various utsavas were celebrated.

The River Ganges which flows through northern India like wise is referred to as a goddess originating from the top of Siva's head in the Himalaya mountains, giving sustenance to hundreds of millions of modern Indians. Modern industrial contaminants and human wastes have fouled the rivers, though Ganges water still plays an important role in India's ritual life.

To Hindus, the concept of environment protection is not a modern phenomenon; they inherited it from their ancestors. During the earliest, formative period of their society, Hindus first perceived God's presence around them through nature. The natural forces that governed their daily lives were considered as manifestations of an almighty creator they called the Brahman. So to please God, they felt that one should live in harmony with his creation including earth, rivers, forests, sun, air and mountain. This belief spawned many rituals that are still followed by traditional Hindus in India. For example, before the foundation of a building is dug, a priest is invited to perform the Bhoomi pooja (a ritual to the Mother Earth) in order to worship and appease mother earth and seek forgiveness for violating her.

The Indian environment-protection movement has failed to become a mass movement for it is devoid of spiritual foundation or content necessary to inspire Hindus. Prof. David Frawley, laments; "Unfortunately, Hindus have forgotten this vedic view of the earth and don't protect their natural environment. They have not added a Hindu point of view to the ecology movement which is perhaps the main idealistic movement in the world today part of the challenge of the modern Hinduism is to reclaim his connection to the earth". The Indian temporal exercise construct their hands to achieve "Dharma, Artha, Kama, and Moksha", the ultimate of life by adopting and worshipping "Vastu". In other words with out 'Vastu' the architectural art has little significance.

In India construction of buildings is actually a religious rite providing four aesthetic achievements of spiritual Bliss, Artha, Desire, Pleasure and Solution. In ashramas of the Rishis, the tiger, goat, snake, mongoose, etc., used to sit side by side forgetting their celestial enemity. After the completion of the building all religious rituals as Vastu cermony, recitation of mantras Gayathri, Rudra, etc., worship of lord Vishnu, Ganesh and nine planets for planetary peace be performed in accordance with the established practice. Feeding the pigeons with gram, giving water to birds, bread to cows and dogs, plants watering the trees without any expectation of reward, gratifying the Brahmins by meals are important features of Hindu culture.

There were so many folk gods like Air (Marut), Sun, Indra, God of Water (Varuna), Dawn (Usha), Rudra (Shankara) etc. "Indra Mahotsava" is celebrated to express pleasure and joy obtained by greenery. In ancient times so many festivals, functions, ceremonies, sports, seminars, and other social ceremonies named 'Soorvasantak' and 'Madanotsava' were very famous which were celebrated with gusto by the Aryans. On very auspicious occasions, the buildings were decorated by ornamental leaves known as "Toran" and "Vandandvaar".

In several castes celebration of marriage between the Ficus and Neem was prevalent as they bestow long life good luck and health. Hindus perform marriage of Tulasi plant with the faith that a child will bless them and it is the most auspicious plant. According to Hindu civilization plantation is an auspicious ritual. They accept that "life" persists in the plant Kingdom and they have the power of perception, learning realization and understanding. As per direction the time of plantation of trees was done according to Muhurt (auspicious time). The inspiration, tree gives shows its spiritual nature, symboling its spiritual power, beauty and longevity.

Trees of Worship in Various Practices

Sami is worshipped on Vijayadasami in almost all parts of Bharat. Wood of Shami is proved to be the best and leaves are used in Yagna. After the second day of Dulhandi (Holi) a religious rite called 'Dashamata' is observed for continuous ten days during which the Peepal tree is worshipped. ARK is considered as a divine herb. Just as sun it is also nurture, protector and well-wisher. 'White Ark' is much more important as Ganapathi is made and worshipped from its roots. Bilva means, "that which destroys disease". It is blessed by nature and given to mankind for its beneficence.

The fruit has arisen from the breast of Lakshmiji, and is an extremely dear tree of Mahadevi. Bilva is worshipped as it is believed that the Lord Brahma, Vishnu, Shiva resides in the root, central position and foremost part respectively. Coconut plays an indispensable role. The dried coconut called Sriphal is offered to deities as it gives Ridhi Siddhi.

Papaya is nick named as 'Madhukarkati' and 'Vat Kumphal'. Lemon and Chillies are tied at the entrance, as it is believed that they prevent an evil eye. The humble neem tree with remarkable curative properties from India enjoys international status today. The idols of the gods in the temple at Jagannath Puri are carved out of neem wood. The practice of having a meal under the amla tree 'Vanabhojan' during Karteeka Masam (November) gives exposure to the individual as it balances the imbalances of the atmosphere at that particular time. With thick bark of Kalpavriksha (Seemachinta) people build houses, which are as good as air-conditioned. Agnihotra and homa also result in the process of atmospheric balance and therapeutic advantages.

Jainism

The goal within Jainism is to ascend to the siddha loka, a world beyond heaven and earth, where all the liberated souls dwell eternally in a state of energy, consciousness, and bliss. Jainism holds some interesting potential for ecological thinking, though its final goal transcends earthly concerns.

At the core of Jaina faith lies five vows that dictate the daily life of Jaina lay persons, monks, and nuns. These five vows, which inspired and influenced Mahatma Gandhi, are non-violence (ahimsa), truthfulness (Satya), non-stealing (Asteya), sexual restraint (brahmacharya), and non-possession (aparigraha). For instance life particles (Jiva) in earth, water, fire, air, micro-organisms and plants each experience the world and they avoid activities associated with violence and follow a vegetarian diet.

Humans have an inescapable ethical responsibility that made the Jain tradition a cradle for the creed of environmental protection and harmony. In its profound sense, the Jain philosophy studies the dualism between matter and spirit and the relationship of the soul to the universe. An important constituent of Jain philosophy is Doctrine of multiple concepts (Anekantavada), since spiritual practices are multifaceted, different standpoints can hold true.

Buddhism

The Buddhist tradition spans 2,500 years and more than a dozen cultures that address important aspects of human nature relations. Contemporary Buddhists are exploring affinities between cardinal Buddhist teachings and ecological philosophies of life that lead to old and new eco-Buddhism continues to develop.

Maharatnakuta Sutra, "Dwelling in the Forest" is a spiritual path related to environment. In Buddhism right views, right action, right livelihood, right mindfulness, right effort, etc. bring out ecological consciousness. Lord Buddha's ethical and ecological visions have compelling relevance for our own time.

Taoism

According to the Taoist Chinese Philosophy of life, all natural phenomena are the manifestations of the perpetual fluctuations between the poles, where all transformations occur gradually in an integrated progression of ecological balance.

Islam

In Islam the role of human being is not only to enjoy, use and benefits from their surroundings, but also preserve, protect and promote their fellow creatures. There is a strong sense of goodness and purity of the earth. Clean dust may be used for ablutions before prayer if clean water is not available. The prophet Muhammad said, "The Earth has been created for me as a mosque and as a means of purification". A religious practice, in which green colour flags are tied to trees, indicates global environmental nature as God's perfect plan.

Shintoism

Shinto is Japan's indigenous religion. It refers to diverse and localized religious beliefs, ritual practices and institutions. Discretion towards nature and Kami (deities) is essential since they nurture human life and believe that the natural and social environments are inter related in spatial and typographical terms. They are manifested in the arrangement of traditional residences in relation to fields, mountains and rivers. Purification (haree) is performed to reestablish order and balance between natures. Harm done or accreted pollution can be neutralised by means of ritual practices.

Judeo-Christianity

The followers of Judaism explore the relationship with the rest of God's creation. The endeavour to formulate a systematic environmental ethic is quite new to Judaism. According to systematic environmental ethic, humanity is more or less a divine after thought in this dynamic picture of God's world. The earth is the Lord's. Observance of sabbath was one of the important spiritual practice. Later this was applied to agricultural policy as its a sound soil conservation policy. Jesus used water, soil, the air, and the fruitfulness of the earth to call attention to indications of the Reign of God breaking in among us. Sacramental symbols come from nature: water, bread, wine, oil, ashes, fire, and the human touch.

Thomas Berry, the cultural scholar and Catholic priest who calls himself a "geologian" rather than a theologian chosen. "The ten commandments may not be completely or adequately represent ecological spirituality, but they are frequently discussed from "Deep Ecology" movement, to Traditional religions and their spiritual practices have hampered or have facilitated the realization of ecological spirituality according to ten commandments. Here Moses does not go up Mount Sinai to receive the tablets from a sky god, but rather he descends into the depths of the earth to discover the new commandments.

1. The universe, our solar system and the earth, as well as our human evolutionary emergence from animal ancestors on this planet constitute the primary sources of revelation of the ultimate mystery.
2. The universe is a unity of matter and energy; it is an interacting community of systems; in its earthly dimension, the psychic and the physical are intimately integrated and operate according to laws of differentiation, subjectivity and communion.
3. A main task for humans is to assist the intercommunication of living and non-living components of the earth community. This involves moving from an exclusively anthropocentric to an organic perspective, one that appreciates the intrinsic not just the instrumental value of nonhuman reality. This requires a profound reorientation among people toward an integrated human-earth relationship.
4. The primordial components of earth: land, air and water are sacred.
5. The richness and diversity of all life forms must be preserved in a way that upholds ecojustice; the expansion of human population and its interference in nature is excessive.
6. People must rethink their consumer habits and move toward styles of simpler living to preserve the earth and establish more enhancing forms of community life.
7. Humans need to re-learn ways of communicating with nature via dialogue and not coercion, thus recovering their true relationship with the life of earth.
8. A new ecological ethic needs to be founded on a deep sense of bonding with nature, as a basically aesthetic experience. Ethical principle and applications can be positively influenced by a prior affective knowing of the natural world.
9. Ecofeminism provides a negative critique of patriarchal structures that have oppressed both women and the environment; this movement also offers positive insights from the experiences of women and nature to enhance ecological spirituality.
10. Humans must learn to relate to the animal world in ways that lessen cruelty and violence, while enhancing inter species relationships with animals that benefit the whole biotic community.

The ten commandments of ecological spirituality summon traditional religions to a profound re-interpretation of our doctrines and practices to encompass the well-being and sustainability of all the biotic systems of this beautiful planet.

Conclusion

Scientists are in general agreement about global warming; expanding populations strain and deplete finite resources; air, land, water etc., polluted by chemicals and radioactive waste; consumerist market systems plunge ahead of long term consequences to biotic systems. Scientific study, governmental policies and public education are crucial elements for addressing these problems. But the environmental challenge today is so great that it calls for a revised human consciousness of what it means to be a creature of earth. This demands a new spiritual understanding and experience of bonding with nature. It is what the socio-biologist, E.O. Wilson terms "biophilia", an attitude of profound respect for and attachment to our natural surroundings.

Spiritual practices have always been rooted in the simplest, most basic things in our lives. To be on the path of sacred ecology today means to take our spiritual work back into the realms of daily practice. The time has come to reverse the course or to go from environmental crisis to spirituality, from awareness of ecological abyss to awareness of the effects of lack of spirituality, so we must move along this path, not by small steps but by giant leaps, in order to avoid dragging on this righteous undertaking over dozens of years. Ecological and cultural restoration then is the same.

The thought about the 'cultural restoration' in to our educational system is the bringing back of those native life ways which contribute to spiritual and material wholeness that maintain our relationship with the family of ecological life is the gateway to the 'Future'.

References

Belopopsky, Alexander and Dimitrioikonomeru, (2005) (Ed.), The Orthodoxy and Ecology Resource Book, Syndesmos, The World Fellowship Orthodox Youth.

Dwivedi, Bhoj Raj, *Environmental Vastu*, New Delhi, Diamond Pocket Books (P) Ltd.

Eagle, Booke Medicine, *The Path of Sacred Ecology*, Rocky Mountain Spiritual Emergence Net Work Times, Vol. 5, Nov.2 (an interview with the author).

Fritz of Capra (1984), *The Tao of Physics*, Bantam Book, Shambhala Publications Inc.

Ignacimuthu, S.J. (1998), *Environmental Awareness and Protection*, New Delhi, Phoenix Publishing House.

MallSinghvi, Lakshmi H.E., *The East is Green*.

Reddy, Narayana (1990), *Acharalu-Sastriyata*, Vijayawada, Sri Gayathri Press.

Richey, Jeff (2002), *Eastern Ecology Free of Western Imposition*.

Sharma, R.A. (2001), *Environmental Education*, Meerut, Surya Publication.

Vedavyas, (1993), *Hinduism in the Space Age*, University of Vedic Sciences, Yoga Brotherhood of America (Inc.).

<www.canfeinesharim.org>

<www.environment.harvard.edn/religion>
<www.biol.tsukuba.ac.jp/umacor/index>
<www.big_picture.tv/transcripts/elisabet sabtouris.pdfratiorsnote>
<www.gdufour@global communitywebnet.com>
<www.boloji.com>
<www.indiatogether.org>
<www.pbmoyle@ucdavis.edu>
<www.releb@emory.edu>
<www.vedicworld.org>

FOREST SPLENDOUR

Marcel Ekka S. J.

As far as the general symbolism of the earth's landscape is concerned, forests occupy a notable place, and are often mentioned in myths, legends and folktales in all the world cultures and civilizations. Forest symbolism is complex, but it is connected at all levels with the symbolism of the female principle, or the Great Mother. The forest is the place where vegetation thrives and luxuriates, free from any control or cultivation. Since its foliage obscures the light of the sun, it is regarded as an opponent to the sun's power and as a symbol of the earth.[1] In Druid mythology, the forest was given to the sun in marriage.[2]

Heinrich Zimmer says that, in contrast with the city, the house and cultivated land, which are all safe areas, the forest harbours all kinds of dangers and demons, enemies and diseases.[3] This is why forests were the first places in nature to be dedicated to the cult of the gods, and why propitiatory offerings were suspended from trees—the tree being, in this case, the equivalent of a sacrificial stake.[4]

Ranchor Prime, born in Leeds, became a devotee of Lord Krishna in Britain and India. He studied in depth the environmental values found in the Hindu traditions, while claiming legitimacy of the *Sanatana Dharma*, which from his perspective roughly means, 'the eternal essence of life.' Moreover, this essence is not only limited to human beings but it also includes animals and plants—with the universe that surrounds them—with the original source (esse) of their existence, the Godhead.

Being a devotee of Lord Krishna, he visited *Vrindavan*, interacted over there with a host of well versed persons of Hindu tradition, who were themselves devotees and had retired from active household obligations to devote to their religion. *Vrindavan* is one of the six forests in the *Vrijbhumi*, which were once the home of Krishna and his cows and the *Gopis*. The devotees of Krishna tutored Ranchor univocally into some of the basic concepts of the environmental balance as found in the traditional pattern India as it was then entirely Hindu. The human settlements have traditionally been between the forest and a pond or lake or river.

The splendour of the moon, the stars, the rising sun, the winds, the vegetation, the animals, rivers, trees and mountains together constitute the beauty of the natural creation in its entirety. Part of that beauty is called *vanavaibhava* (forest splendour). Human beings are part of this forest splendour and should therefore, in their natural state love and respect it. We are part of it, it gives us our identity, and without it we are lost. It is mandatory for us to carefully nourish and preserve the forest.[5]

The first type of forest in the traditional pattern of Hindu India was called *raksha* (sanctuary). It would be left entirely to itself and no human would enter it. It was completely protected from human disturbances. During the daytime the birds and the animals would go into the village or wherever they wished but could return safely to their habitat at night. All of them living in that patch would feel quite safe. The *raksha* forest corresponds to both the *mahavan* and *tapovan*.[6]

The *shrivan,* the second type of forest has two features—*ghana* (dense) woodland and *vanakhandi* (groves). In this forest the natural arrangement of trees and plants would not be unnecessarily disturbed, but villagers could go there to collect dry wood, leaves, forest produce and a limited amount of green timber. The village communities looked after it as their livelihood depended on it; they naturally conserved it from generation to generation.[7]

The third type of forest was called *vanakhandi* (groves). It is usually of one particular species, specially planted by the villagers—*amara-khandi* (mango), *kadamba-khandi, tamal-khandi* etc. The groves usually comprised of fruit trees and were maintained by the villagers for religious, social and recreational purposes. Most recreations, such as dancing and singing had to do with religious festivals like *raslila,* the circle dance of Krishna. This commemorates Sri Krishna's dancing with the *gopis* (cowherd girls) during the full moon night of the autumn season in the sacred groves on the bank of the river Yamuna.[8]

What is the religious significance of a forest? The *tapovan* was the home of the sages. *Tapa* means penance and *vana* means forest. The sage or *rishi* must live in a place that gives the ambience of the presence of God—this is the *tapovan*. If one wished to meet such advanced souls, one had to go to the *tapovan* where their hermitage or ashrams could be found. There are many stories in *Vedic* literature of encounters between worldly persons and sages in their forest ashrams. From this profoundly natural setting emerged the *Vedic* teachings of the *Upanishads* such as the *Brihad-Aranyaka,* which means the teaching, which began in the forest.[9]

To understand our relationship with *vanvaibhava* (forest splendour), we must look within ourselves and thus begin the spiritual path: spirituality begins from looking within. True spirituality means looking not only within us, but also the whole of nature—to understand the internal reality of nature. Culture is the outward expression of spirituality. From human culture comes heritage—the permanent expressions left in stone, art and literature—those things by which we pass on our values to the next generation. To know these cultures and heritages we must travel to different sacred places to find out how all these different patterns are tied together—to find unity in diversity. To go on pilgrimage is to experience the values of a particular place, to feel one with the culture of that place.[10]

Mahatma Gandhi has said that of all the animal creations of God, man is the only animal who has been created in order that he may know his Maker. Man's aim in life is not, therefore, to add from day to day his material prospects, and his material possessions, but his predominant calling from day to day is to come to his own Maker. In the entire biosphere, man is the pinnacle of creation. Merely being born a human being, committing wanton destruction upon the environment and amassing wealth does not entitle one to be called a civilized human being. Gandhi gave humankind the role of a trustee of all other living creatures. Therefore, the human race has to earn this distinction by better thoughts and deeds.[11] Being kind to others includes being kind to nature. Nature's welfare and human welfare cannot be separated from each other. For this reason Vedic culture taught that the earth and the cow are to be loved and cared for as mothers. If we are kind to nature we will naturally be kind to one another. In our Indian perception, *manav* is a human being who perfectly respects nature and *danav* is one who misuses nature. It is not wise to go against nature.[12]

Consistent over cropping of any species can reduce its number to levels from which recovery is no longer possible. Killing the last animal or cutting the last tree of a species may not be needed to bring about ex-termination. If animal populations are reduced to a level at which effective breeding does not occur, or if the habitat is destroyed, recovery will not take place, even though some individuals may for a long time survive. Similarly for plants, if a minimum area to ensure regeneration of certain species is not maintained, they will eventually die out.[13]

In the history of America, the frontier view of nature as an object of plunder without defences has led to environmental depletion. One time even American public law offered inducement to exploitation of natural resources under the doctrines of the sanctity of private property. Virtually all instances of environmental pollution stem from profligate enjoyment by a private individual of his property right. Much of the same bias has been retained in contemporary legal doctrine. The same legal doctrine, which once justified, interpreted and safeguarded prodigality in the use of natural resources, which led to the forest denudation, can be a potentially effective forum for reversing the trend of history.[14]

Modern man has initiated a number of abrupt changes beyond the capacity of the ecosystem to adjust himself. He has become the dominant organism in the biosphere, forcing changes to fit his needs and desires, and creating ecosystems different from any that have existed before. This acquisition of dominance is the result of the extraordinary evaluation in man of knowledge, ingenuity, and use of tools. He has greatly modified or replaced the natural processes that control stability and balance within the ecosystem with new processes and manufactured products, which the environment cannot absorb. The result has been a general deterioration of the environment, a population explosion of the human species, and a disharmony in the world ecosystem unless resolved, will quickly lead to catastrophe. Whether man will survive as the dominant organism, be reduced again to the role he occupied as a primitive, or become extinct may well be decided within the next few decades.[15]

There is strong evidence that thick forests once covered India. Civilization flourished in the sub-continent way back in 2000 B. C. In the early years, the Aryans cleared the forests for agriculture. They maintained all their big institutions in the sylvan surroundings of the forests. Attractive descriptions of forests like *Dandakaranya* and *Nandanvan* find mention in the epics *Mahabharata* and *Ramayana*. The people by and large held forests in great reverence. Groves were raised near the temples.[16]

Alexander the Great invaded the Indian sub-continent in 326 B.C. It was the first time when the countries of the western world got an account of the physiography and forests of India. Historically, the first Indian ruler, who recognized the importance of forests, was Chandra Gupta Maurya, who came to power in 323 B.C. after the departure of Alexander. He appointed *Kupyadhyaksha* (Manager of the Forests) to look after the forests of his kingdom. This high official was assisted by a number of subordinates.[17]

During the rule of Emperor Ashoka, forests and forestry assumed a new dimension owing to his conversion to Buddhism. The *ahimsa* (non-violence) characteristic of his rule was reflected not only towards animals but also towards the trees and plants. Emphasis was laid on the planting of trees, along the roads and paths and camping sites. Forest sites like *Lumbinivan* and *Uruvela* connected to the life of Lord Buddha enjoyed royal patronage. According to *Divyaudan* (Buddhist literature), Ashoka made a grant of one-lakh gold coins for the protection, development and regeneration of trees of the *Lumbinivan*.[18]

During the Medieval Age, the early Muslim invaders did not pay attention to the conservation of forests. Their invasion uprooted many people from their homes. The vanquished sought refuge in forests, which were cleared for new settlements. The Mughal Emperor, Akbar took keen interest in planting trees along the roads and canals. Many gardens were built known as Mughal gardens. After the downfall of the Mughal Empire, the Provincial Kingdoms frequently fought each other. As a result, a large population abandoned cultivation and moved to safer places, where forests were cleared for settlements.[19]

The British regime in India exploited unabated the timber wealth of the country. The sub-continent abounded in forest resources so much that the British considered them to be inexhaustible. The general policy was to clear forests for agriculture and to use timber for shipbuilding and general construction. As a result, the need for regulated exploitation of forest sources was felt. The Governor General Lord Dalhousie promulgated for the first time in 1855 an outline for forest conservancy for the whole of the country. In the Presidencies and British ruled Provinces, steps were taken to prevent uncontrolled destruction of the forests and their over-exploitation.[20]

The First World War once again saw severe depletion of India's forest resources. Vast quantities of timber were transported for shipbuilding and railway lines and to pay for the British war efforts. After the end of the First World War, the history of Indian forestry passed into the next phase (1919-1947). In the political changes of the twenties, forests were transferred to the jurisdiction of the Provincial Governments. In 1926, training of Superior Forest Service Officers began at Dehradun. The Provinces maintained their own separate cadres for the higher level of officers. Afforestation works were taken up on a large scale in many parts of the country. After the end of the Second World War, a number of forestry schemes were initiated, which included large scale plantation activity, setting up of forest based industries, revision of working plans and thrust on research.[21]

Post-independent India witnessed rapid development of forestry. The landmarks of Indian forestry during this period are: Setting up of the Central Forest Commission, Formulation of the National Forest Policy in 1952 and 1988, Launching of social forestry in a big way, Stress on wildlife conservation through biosphere reserves, national parks, sanctuaries and tiger reserves, Enactment of the Forest Conservation Act in 1980, Creation of the Indian Council of Forest Research and Education and thrust on forest conservation and rehabilitation.[22]

After having sung praises to the forest splendour and tracing the history of forestry in our Indian context, what is alarming, is the fast pace of deforestation. *Forests precede civilizations, deserts follow them,* so said Francois-Rene de Chateaubriand. We are all aware that the lives and livelihood of half the world's population depends directly on the forests and ecosystems. In recent years, deforestation has had an impact on the climatic conditions causing global warming. Even *Vrindavan,* once famous for its lush green sacred groves are almost denuded. The cruel hands of the present day landscape developers have been scrapping the last vestiges of the remaining trees in *Vrindavan*. A Krishna devotee, Ranchor Prime emphatically says, "the importance of trees to Krishna and the whole awareness of Krishna in *Vrindavan* is fundamental."[23] Obviously, the conclusion then is "no Nature/Forest, no God experience."

All of life, from the universe itself down to the individual trees and seed and the very earth beneath our feet, is full of the divine purpose of God behind it. Unless we understand this divine purpose, which lies behind the world, we will not know how to live in it properly, or how to use it. Science has greatly advanced in the last two hundred years, making it possible to manipulate nature in ways that were previously impossible. This has brought many benefits such as medicines and comforts of life, but it has also created many dangers. If we do not know what the world is meant for, if we do not know how to use it properly, we can do immense harm with the power that modern science has given us. Therefore, the Hindu Scriptures have advised that the knowledge of matter, namely science, must be coupled with the knowledge of the spirit if it is to be of benefit to humanity.[24]

As human beings, our first contact is with the natural environment of the forest splendour. To realize the divine, we just cannot bypass nature. If there is any wisdom in today's context, it is ecological wisdom. The ecological wisdom should be incorporated in all our decision-making processes in order to save the web of life. The modern world needs more and more nature mystics. "Religion that sets itself in opposition to nature and vehemently resists the resacralizing of the natural world on the grounds that it would be pagan to do so is not sustainable over time."[25]

Notes

1. J.E. Cirlot, *A Dictionary of Symbols*, Routledge & Kegan Paul, London, 1981, p. 12.
2. *Ibid*. A Druid was a member of a learned class of priests, teachers, and judges among the Celtic peoples. The Druids instructed young men, oversaw sacrifices, judged quarrels and decreed penalties. They were exempted from warfare and from paying tribute. Cf. *Britannica Readymade Encyclopaedia*, Encyclopaedia Britannica (India) Pvt. Ltd., New Delhi, 2005, pp. 187-188.
3. *Ibid*. Heinrich Zimmer authored *Kunstform und Yoga im Indischen Kultbild*, Berlin, 1926, in which he mentions negative symbolism of forests from cultic perspective and not otherwise.
4. *Ibid*.
5. Ranchor Prime, *Hinduism and Ecology*, Motilal Banarasidass Publishers Private Limited, Delhi, Reprint Edition, 1996, pp. 19-20. A Krishna devotee, Sri Sewak Saran, at *Vrindavan* explains the meaning of *vanavaibhava* to Ranchor.
6. *Ibid*. p. 17.
7. *Ibid*.
8. *Ibid*. p. 18.
9. *Ibid*. p. 12. Ranchor met Sri Banwari, a devotee of Krishna at *Vrindavan*. The explanation of *tapovan* is that of Banwari.
10. *Ibid*. pp. 20-21.

11. Nature and Humankind taken from Mahatma Gandhi: An Apostle of Applied Human Ecology, p. 18.
12. Ranchor Prime, pp. 21-22.
13. Aaradhana P.S., *Community Ecology*, Rajat Publication, New Delhi, 1998, p. 38.
14. Gerald Garvey, *Energy, Ecology, Economy*, W. W. Norton and Company, Inc. New York, 1792, p. 198. To subdue the material potential became the dominant aspiration of American frontiersmen. The seemingly limitless bounty of the American continent fostered reckless wasteful habits of natural resources.
15. Aaradhana P.S., p. 39.
16. S.S. Negi, *India's Forests, Forestry and Wildlife*, Indus Publishing Company, New Delhi, 1994, p. 21.
17. *Ibid.* Cf. K.L. Khurana, *History of India*, Lakshmi Narayan Agarwal, Agra, 2001, p. 139.
18. Goyal Shriram, *Praachin Bhartiya Abhilekha Sangraha*, p. 118. Quoted by Ajay Kumar Singh in his M. Phil. Dissertation (unpublished) entitled, *"Praachin Bhartiya Abhilekhon mei varnit Nirmaan Kaaryon ka Adhyayan"*, Rani Durgawati Vishwavidyalaya, Jabalpur, 2005.
19. S.S. Negi, pp. 21-22.
20. *Ibid.* pp. 23-24.
21. *Ibid.* pp. 26-27.
22. *Ibid.* p. 27.
23. Ranchor Prime, p. 109.
24. *Ibid.* pp. 3-4.
25. Carolyn Merchant (Editor), *Ecology*, Rawat Publications, New Delhi, 1996. p. 299, from *"The Spiritual Dimension of Green Politics"* by Charlene Spretnak.

ECOLOGY IN HINDUISM AND ISLAM

Gargi Mukhopadhyay

India has a rich and varied religious tradition. The unity amidst religious diversity presents India with a pride of place among all the countries of the world. India's long secular traditions nurtured many religions and provided conditions for them to come to their full bloom. The country experienced rare moments of glory with the emergence of famous religious preachers, religious reformers, and saints and held high their gospels of humanity, fraternity and the doctrine of 'live and let live'.

All religions provide a special place to the environment around us, the PRAKRITI (nature) as it is called in Sanskrit, which reveals itself in the green foliage, flowers, fruits, rivers, springs, mountains, the sun, the moon, the stars, the air around us and also the plant and animal kingdom. This beautiful, awe striking, sometimes silent and sometimes violent nature has always evoked a sense of respect, awe and wonder in our minds. While on the one hand nature mesmerizes us with its exquisite beauty, on the other hand its furies make us helpless. That is the reason early men started worshipping the furies of nature to please them. With the passage of time, these rituals for keeping the furies of nature at bay secured newer interpretations for themselves; interpretations that made nature meaningful to us. Ecology thus, entered the religious scriptures and blended itself with the faith in God, with whom we associate the creation of this Universe.

The present paper tries to point out the similarities of interpretations between Hinduism and Islam, in regard to the relationship of ecology and religion. The interpretations of these two religions in regard to ecology reveal striking similarities stressing the importance of ecology in the life of human beings. Both appeal to humankind to show concern and care and try to protect the precious environment around us of which we are also a part and parcel.

Evolved and nurtured in two completely different cultural milieus both Islam and Hinduism look at the relationship between ecology and humankind from individual viewpoints. While the Hindu interpretations of ecology reveals itself through personification of the various components of the universe or associating rituals with one factor or the other, Islam keeps faith in one single Creator, Allah, the Almighty, and tries to make humankind aware of their debts and duties towards their environment. Both the religions have given vent to their feelings in exquisite slokas or hymns, be it the Vedas, the Upanishads, and the Puranas, or the Holy Quran.

Hinduism has always tried to see and realize the greatness of God amidst his creations. If we take a keen look at the rituals that guide the life of a Hindu from the cradle to the pyre the implications are easily identified. The denouncement of material pleasures and the respect, surrender and total devotion to God and his creations, has been identified as the path towards mukti or salvation, which should be the ultimate goal of humankind. Human beings are directed to do satkarmas or pious duties towards the living creatures and the environment, which sustains life. Only and only then can peace and happiness be realized. According to the Hindu scriptures "true happiness comes from within and not from outer possessions. Material possessions should never be allowed to dominate over one's life. Life's main purpose is to discover the spiritual nature and search for the peace and fulfillment it brings."[1]

Very much in the same tone Islam speaks about the supreme Creator, Allah, the Almighty, who according to them is the Creator of this Universe. This humankind and the environment consisting of the air, water, fire; soil, plants and animals owe their existence to him. All his creations must be taken care of. The Holy Quran awards the responsibilities of protecting the Universe on humankind who will always strive to work for the betterment of their Universe as the chosen deputies of the Almighty. Man cannot live all by himself and hence we should and must remain thankful to Allah and his creations without which our very existence would have been at stake. The sky, earth, air, water and fire are considered as the five major elements, which sustain life on earth. Hinduism worships these five elements as "Panchabhuta" who works in accordance with the laws of nature. The Vedas explain that, all material bodies are composed of these elements and disintegrate into these elements after death. Not only human beings but the plants and animals too owe their existence to them.

Fazlun Khalid, in his article, "Guardians of the Natural Order", points out that the whole humankind, in their bid to attain progress and development, have completely shifted their attention from the environment. Islam believes in Allah, the Almighty, as the sole Creator of this Universe and the celestial bodies, the plants, animals and humans; all obey his commands. They support one another to sustain life on this Universe. According to the Holy Quran: *"To Him belongs whatsoever is in the heavens and the Earth, all obey his will and it is He who originates Creation...."30:25*[2]

The Earth also has a place of pride in the Islamic as well as the Hindu faiths. It is on this soil that vegetation springs, crops are grown, and men raise civilizations. In Hindu faith just as the mother cares for the children the earth holds life. Islam also has used the mother and child concept in regard to the relation between human beings and the earth. The Holy Quran urges its followers to respect Earth as the Mother and remain thankful to her for her services towards mankind. Fazlun Khalid states that "Islamic teachings offer an opportunity to understand the natural order and to define human responsibility."[3]

Special importance has been laid upon conservation and preservation of water; another most important element responsible for life on earth in Islam. It regards water as the very lifeblood of this Universe and "forbids the wastage of water and the usage thereof without benefit."[4] The holy Quran regards the preservation of water as a duty for the "drinking of mankind, animal life, bird life, and vegetation"[5] the proper performance of which will please Allah. Human beings are also forbidden to pollute or contaminate the water bodies thereby trying to prevent harm being extended to the aquatic plants and animals.

Water has been renamed "jeevan" meaning life in Hinduism. Not only in religious texts but also in literatures, water has been praised as "amrit" or the heavenly drink that quenches thirst and soothes the soul. As far as the importance of rivers is concerned they occupy a central place in Hindu rituals and culture. Hymns are found in the Vedas praising rivers, especially the Ganga (Ganges), who is believed to have the capacity of washing away the sins of the individuals. The Indian rivers have always been considered sacred by the Hindus and pollution of these waterways is considered as heinous sins. Hinduism urges the human race to work towards purification of these water bodies and help save underwater lives.

The plant and animal world is given a high and respectful position in the Hindu religious faith because many of the plants and trees are worshipped as incarnations of various Gods and Goddesses while a variety of birds and animals are found to accompany these deities as their carriers or Vahanas. Thus the peepal tree, the banyan tree, and a very many medicinal plants and the Hindus with great devotion worship herbs. Among the animals receiving attention and worship are: the Lion as the vahana of Goddess Durga, the Bull as the vahana of Lord Shiva, the Peacock as the vahana of Kartikeya, the Commander-in-chief of the Heavenly army, the Rat as the vahana of Lord Ganesha, and so on. The inner implication is to preserve and protect the varied ness of the plant and animal kingdom and to make human beings realize their importance.

Islam stands by the plant and animal kingdom, and requests human beings to show kindness and provide protection to them. According to Islam, "we must deal with them with utmost beneficence and compassion, and strive to ensure the preservations of different species of animal life."[6]

The Holy Quran has also warned us not to harm or kill birds and animals for mere amusements. "Islam has forbidden wastage of animals and plants in peacetime and wartime."[7]

Both Hinduism and Islam have recognized the role of natural vegetation in their scriptures and teachings. The greenery that covers the earth's surface not only helps sustain the nitrogenous cycle but also provides oxygen, which we, humans, breathe. The balance is maintained in the way that the carbon dioxide, which we exhale, is the source of life of the plants and animals. That is the equilibrium which nature maintains. Also the exotic and beautiful nature, with its serenity and vastness helps us realize deeply the Ananda or true pleasure and gives us a taste of Mukti from the material cycle of pleasures and pains. The Hindu as well as the Islamic faith believes that nature is by far one of the best creations of God who has taken great care and concern in creating it. The colors in the petals and on the wings of the butterflies, the rippling rivers with the sunlight reflecting and breaking with the ripples, the chirping of birds, all point to the fact that God loves nature the most, says Hinduism. This view has been expressed in the Holy Quran as well.

"The Exalted says, *He it is who sends down water from the sky, and therewith we bring forth buds of every kind; we bring forth the green blade from which we bring forth the thick clustered grain; and from the date palm, from the pollen thereof, spring pendant bunches and gardens of grapes and the olive and the pomegranate, alike and unlike. Look upon the fruit thereof, when they bear fruit thereof and upon it's ripening. Lo! Herein verily are signs for a people who believe." (Ch.6; v.99)* [8]

For Hinduism, Prakriti or Nature is incomplete without Purusha, the Humanity and vice versa. One cannot sustain without the other. Thus, human beings and his environment are interdependent; they can survive only in unison. Demolition of the environment around us spells death and destruction of all living creatures, a fact which none can deny. The two religions beautifully clarify the idea of "live and let live" which is central to the fact of existence. Men must realize that they cannot sustain themselves without the environment. In his bid to attain development he has destroyed and is still destroying the environment robbing it of peace and serenity. Life is meant to be shared. And they care for you, for whom you care; this is a give and take relationship where the natural law perfectly balances every living creature with the other. This simple give and take relationship is the keyword for healthy living. Man must stand by his environment not as an enemy but as a friend, protector and preserver. Only then can he fulfill the purpose of his being the most special among God's creations.

According to Shrivastava Goswami, (Vaishnava Acharya of Shri Radharaman Temple, Vrindavan; Chairman of the Vrindavan Conservation Project): *"This awareness of ecological play or playful ecology is inseparable from awareness or the need for friendship and play as the real basis for human relationship. The family within which these relationships are nourished is not limited to its human members. Just as the human child has to be nourished by Mother Nature, and the human spirit has to be embraced and loved by beautiful nature, so the human being who has grown old or sick has to be supported by caring nature. If humans distress the Mother, rape the beauty, and beat the caring nurse, what happens? The relationship collapses and the families are broken."*[9]

Notes

1. What Does Hinduism Teach About Ecology: Hindu Faith Statement in <http://www.arcworld.org/faiths.asp?pageID=43>
2. Fazlun Khalid: Guardians of the Natural Order, in Culture, in Our Planet 8.2 –August 1998, in http://www.ourplanet.com/txtversn/82/khalid.html.
3. *Ibid.*
4. The Water; in Islam and Ecology (Answers provided Imam Tajuddin H Alhilaly, Mufty for Australia(~1993); Translated and some comments added with permission by Keysar Trad) in <http://www.speednet.com.au/~keysar/ecology.htm>
5. *Ibid.*
6. The Animal Kingdom, in Ibid.
7. *Ibid.*
8. *Ibid.*
9. Section from Shrivastava Goswami: Hindu Faith Statement; in What Does Hinduism Teach About Ecology; (This statement edited and compiled by Ranchor Prime, was printed along with statements from ten other faiths in "Faith In Conservation" by Martin Palmer with Victoria Finlay published by the World Bank in 2003); <http://www.arcworld.org/faiths.asp?page ID=43>

RELIGIOUS FAITH TO PRESERVE ECOLOGY

K. Narayanan

The World is in the throes of a formidable ecological crisis due to the reckless exploitation of Nature and all its resources by man. The rich diversity of Nature which has evolved over a period of 3500 million years and which is crucial to man's survival too is at stake. Human development threatens to destroy the kaleidoscopic diversity of Mother Nature in a fraction of the time it took to create. The human centric approach to the exploitation of natural resources of the Western countries based on wrong interpretation of the Old Testament, have taken us to the brink of a natural holocaust. The drive to have mastery over creation has resulted in the senseless exploitation of nature, the alienation of the land from people and the destruction of indigenous culture.[1]

We need to redefine our beliefs and reorient our spirituality. We must realize the purpose of our existence as humans on earth. The fact that we are just one species among millions on a plane which orbits just a medium sized star of the Solar system which is only four and a half billion years, just one among hundred billion galaxies in a Universe that is infinite needs to be noted by us in all humility. Our ancient religions did recognize this reality and have in them religious practices and beliefs, which emphasize the co-existence of all forms of creation. The West had dismissed these as pagan practices and superstitions. The purpose of this paper is to examine as to how a reorientation of our religious beliefs can help us in the preservation of ecology and such an endeavor is most opportune since the world has woken up to the ecological significance of some of the time tested practices and beliefs. The focus of my paper is on 'Hinduism and preservation of ecology with special reference to the Sacred Grooves of India'.

Hinduism has always been proximate to Nature. It calls upon the followers to see god in every creature of the Universe. The Bhagwad Gita quotes Krishna as saying, "I pervade the Universe. All objects in the Universe rest on me as pearls on the thread of a garland." The Upanishad observes, "tat sristvata davanu pravisat" [after creating the universe he entered into every object created]. The Hindus believe that there is soul in all plants and animals and that all forms of life enjoy same importance in the Universe. They are the crucial links in the ecological chain. Like all ancient civilizations of the yore, India has a long history of nature worship and the sacred groves of India are the living testimony of our veneration of nature even to this day.

The sacred groves, which are dedicated to local deities, are protected by local communities through traditions and taboos that incorporate spiritual and ecological values. These have been preserved for generations and they represent the native vegetation in its natural state and are thus rich in biodiversity. The groves are home to many rare species of plants and are the ecological treasures, which are the last refuge for endemic and endangered plants and animal species. Many groves have water bodies like ponds and tanks. The vegetative mass that covers the floor of the groves help in absorbing water during the rainy season. The extensive foliage and tree cover act as a buffer to prevent the run-off of rainwater and this can contribute towards the percolation of water to the underground aquifers. Villages located near large groves have not been affected by water scarcity. A case in study is a 50-hectare stretch of sacred grove in the Ernakulam district of Kerala, which has been zealously protected by the majority Christian community primarily due to its ecological value. It remains as a pristine evergreen forest even to this day. It shows how a common concern for ecology can override even religious differences.

There are more than 15000 sacred groves all over India and 2000 in Kerala.[2] The authors point out that this total "is only an indication of the extent and magnitude of the presence of sacred groves in the country...but in view of the known presence and pattern of distribution of sacred groves in Chhatisgarh, Jharkhand, Orissa, Uttaranchal, Madhya Pradesh and West Bengal, for which detailed inventories are not available, we strongly feel that the number of sacred groves in India is likely to be between 100,000 and 150,000." The authors emphasize the need to inventory sacred groves in every state.

The sacred groves of India are diverse, as diverse as the nation. Some are found linked to the forests of the Western Ghats while others are located in arid deserts of the northwest. Some contain only a few trees while others are hundreds of acres in size. Despite the diversity, the sacred groves have helped in preserving ecological balance and bio-diversity of the respective regions.

Any discussion on sacred groves will remain incomplete unless we acknowledge the unique role of the Bishnoi tribes in managing the sacred groves known as Orans in the desert regions of northwest India. Experts have noted that the Bishnoi belt supports a higher density of human and animal population than any other desert region in the world despite the sparse vegetation and highly limited water resources. This is due to the unique conservation practices of the Bishnois who believe that all living things have a right to live and share resources. They are known for the protection of the Indian gazelle and the black buck. The sacred groves of Rajasthan are believed to exceed thousands of acres.

The Sacred Groves in Perspective

In recent years the sacred groves have been affected by social changes, modernization and large-scale commercialization. The greatest challenge is the loss of traditional wisdom due to the onslaught of the western urban culture. The ecological significance of the sacred groves makes no meaning to most people and their concern is limited to meaningless rituals. Concrete shrines have replaced many groves dedicated to snake worship. The ecological factors have been given to the go by at the altar of blind ritualism. However, many conservationists, NGO's and government organizations have realized that traditional wisdom must be integrated into modern systems. In 1999 the Indira Gandhi Rashtriya Manav Sangrahalaya (the National Museum of Mankind) and several other institutions established a museum for sacred groves with the goal of raising awareness of the important role these ancient forests play in conservation. Located on a 200-acre campus in Bhopal, the outdoor museum replicates plots of sacred groves from different parts of India including the Orans of Rajasthan and the Devarakadus of Karnataka. The museum also celebrates the communities, festivals and rituals associated with each grove. Additionally, the museum has organized a traveling exhibition to build greater ties with local communities and organizations throughout the country, as well as a Sacred Grove Festival.[3]

Earth Watch Institute, in association with Society of Restoration- India has undertaken a sacred grove restoration project in the state of Goa. The project aims at restoring degraded groves; document traditional management practices and educates local communities. Seminars and conferences such as these can go a long way in propagating the spirit behind the sacred groves. The author gained immense insights only after attending a conference organized by the Kerala Chapter of the Association of Christian Higher Education. To sum up the role of religious faith in the preservation of ecology let me quote from the "Orthodox and Ecological Crisis Clause" published by the Orthodox Church in 1990 which uses the image of the priest and the Eucharist to illustrate as to how our destructive power could be converted as a sacrifice for the preservation of the ecology.

"Just as the priest at Eucharist offers the fullness of creation and receives it back as the blessings of grace in the form of consecrated bread and wine, to share with others so we must be the channel through which God's grace and deliverance is shared with all creation. The human being is simply yet gloriously the means for the expression of creation in its fullness and the coming of God's deliverance for all creation."

Notes

1. World Council of Churches at Norwau in 1988.
2. Kailash C. Malhotra, Yogesh Gokhale and Sudipta Chatterjee, *Cultural and Ecological Dimensions of Sacred Groves of India.*
3. An abstract from the website 'Sacred Land Film Project'.

CREATION TO CONSERVATION

J. Manoharan Thomas

God created the heavens and the earth. The earth was without form and void. Then God started creating things one by one, each and every component to form an ecosystem. God's creations comprise all life forms from plant and animal life to microorganism and the water, land, air and light in which they live. In other words it can be stated as the variety of genes, species and ecosystem found on our planet the Earth. The totality of these refers to the term bio-diversity. God saw that it was good when He created the components of ecosystem.

Food and Diet–Vegetarianism (Eco System)

God created the world with all its components. The purpose of creation is that it should survive and multiply. The important thing necessary for survival is energy. For energy all life forms have to consume something. What the living creatures consume is called food or diet. According to Genesis the original diet prescribed for man and beast was vegetarianism. And God said, "See, I have given you every herb that yields seed which is on the face of all the earth and every tree whose fruit yields seed; to you it shall be food".

"Also to every beast of the earth, to every bud of the air, and to everything that creeps on the earth; in which there is life I have given every green herb for food" (Genesis 1:29, 30). So, the living components of ecosystem at the time of creation were the producers, the consumers and decomposers. All the consumers are primary consumer that is pure vegetarians.

Man's Role on the Conservation of Nature

"The Lord God planted a garden, eastwards in Eden and there he put the man whom he had formed" (Genesis 2:8). "Then the Lord God took the man and put him in the Garden of Eden to tend and keep it" (Genesis 2:15). God assigned an important responsibility to Adam. He had to take care of and to look after the Garden of Eden. This implies the responsibility of preserving flora. "Out of the ground the Lord God formed every beast of the field and every bird of the air and brought them to Adam to see what he would call them. And whatever Adam called each living creature that was its name".

"So Adam gave names to Call cattle to the birds of the air and to every beast of the field" (Genesis 2:19, 20). The second important task God assigned to Adam was identification of fauna. So Adam's role was to preserve nature. In other words, he can be called a conservationist. In general he was the chief conservator of forests.

Need for Reformation

"Now it came to pass when man began to multiply on the face of the earth and daughters were born to them that the sons of God saw the daughters of men that they were beautiful". "There were giants on the earth in those days and also afterwards, when the sons of God came in to the daughters of men and they bore children to them" (Genesis 6:1, 2). Now Nephilim or giants, a hybrid of sons of God and human species came into existence out numbering the pure species. Consequently the badness of men was abundant in the earth. Every inclination of the thoughts of men's heart was only bad all the time. The earth was corrupt. All flesh had corrupted their way on the earth. As a result the earth was filled with violence. The creation deviated from its normal course and was gradually losing its originality. God decided to wipe all men and beast from the surface of the earth except Noah, because Noah was a just man, perfect in his generations.

Reformation and Preservation

Before destruction God made suitable arrangements for the preservation of ecological system on earth. God wanted to preserve nature with all its components in tact. So he commanded Noah to make an ark, a shelter for humans and other creatures to escape the flood. "Make yourself an ark of gopher wood; make rooms in the ark and cover it inside and outside with pitch" (Genesis 6:14). Noah did everything the Lord commanded him. He boarded the ark with his wife and sons and their wives to escape the flood. With him were all the various kinds of animals–seven each of every clean animal, a male and his female–two each of animals that are unclean a male and his female. Also seven each of birds of the air-male and female were with him. This arrangement was made by God that there will be every kind of life reproducing again after the flood was ended. "And of every living thing of all flesh you shall bring two of every sort into the ark, to keep them alive with you; they shall be male and female".

"Of the birds after their kind, of animals after their kind and of every creeping thing of the earth after its kind, two of every kind will come to you to keep them alive" (Genesis 6:19, 20). "You shall take with you seven each of every clean animal, a male and his female; two each of animals that are unclean, a male and his female". "Also seven each of birds of the air, male and female to keep the species alive on the face of all the earth" (Genesis 7:2, 3).

Non-Vegetarianism

"And the fear of you and the dread of you shall be on every beast of the earth, on every bird of the air, on all that move on the earth, and on all the fish of the sea. They are given into your hand. "Every moving thing that lives shall be food for you. I have given you all things, even as the green herbs" (Genesis 9:2, 3).

After Reformation

Man was permitted to eat the flesh of animals. This refers to the consumption of all animals resulting in the depletion of the number of all animals both prey and the predators. The predators are termed as unclean and prey as clean. According to Bible the ratio of predators to prey was fixed as 7:2 in the case of animals and later reduced to 6:2. "You shall take with you seven each of every clean animal a male and his female, two each of animals that are unclean a male and his female. "Also seven each of birds of the air, male and female to keep the species alive on the face of all the earth" (Genesis 7:2, 3). "Then Noah built an alter to the Lord and took of every clean animal and of every clean bird and offered burnt offerings on the alter" (Genesis 8:20).

Conservation

The non-vegetarianism permitted humans to consume everything: the prey and the predator. If it continued in course of time there would have been an imbalance between prey and predator. So as a conservation technique dietary law was modified with ban on consuming predators. For the purpose of conservation of animals and birds, God classified them as 'clean' and unclean. Clean birds, animals and insects were generally those that were plant eaters, while those unclean were meat or carrion eaters. Chickens, cattle and grasshoppers were clean while vultures, weasels and flies were unclean. Yet there were some exceptions that may have been due to their being common carriers of deadly bacterial, viral or parasitical infections (e.g.–rat–bubonic plague, bats–rabies, and pigs–trichinosis) Clean sea creatures were apparently those that were found in fresh-flowing water (e.g. bass, trout), while those unclean tended to be bottom-dwellers (e.g. shellfish, catfish) where natural contaminants collected–a factor even more relevant in our modern world where vast quantities of chemical wastes accumulate in the mud of rivers, lakes and oceans. A full explanation and example listing of clean and unclean creatures can be found in Leviticus chapter 11 and Deuteronomy 14:1-21.

Clean For Food	*Unclean For Food*
cattle	snakes, lizards
sheep	camels, horses
goats	weasels, rabbits
chickens	pigs
elk	vultures,ravens, crows,owls
deer, antelope	bats, rats, flies
water creatures that have fins and scales	water creatures that do not have fins and scales

Law of Food

- And you may eat every animal with cloven hooves, having the hoof split into two parts, and that chews the end, among the animals.
- Nevertheless, of those that chew the end or have cloven hooves, you shall not eat, such as these: the camel, the hare, and the rock hyrax, for they chew the end but do not have cloven hooves; they are unclean for you.
- Also the swine is unclean for you, because it has cloven hooves, yet does not chew the end, you shall not eat their flesh or touch their dead carcasses.
- These you may eat of all that are in the waters; you may eat all that have fins and scales.
- And whatever does not have fins and scales you shall not eat; it is unclean for you.
- All clean birds you may eat.

Conservation/Preservation Trees

"When you besiege a city for a long time, while making war against it to take it, you shall not destroy its trees by wielding an axe against them, if you can eat of them, do not cut them down to use in the siege, for the trees of the field is man's food." Only the trees, which you know, are not trees for food you may destroy and cut down, to build sieges works against the city that makes war with you, until it is subdued. (Deuteronomy 20:19, 20).

Later due to the economic and political development nations resorted to war against other countries and tribes. People lived in towns surrounded by fortress. At the time of war to force the people living inside the fortress, the warring nation would raise a siege. Ramming was the method in olden days to smash the door or the weak point in a fortress; for ramming heavy logs were used. So fruit bearing trees can be preserved for the benefit of humans and other birds and animals.

Conclusion

Man enjoys a special high ranking status in the hierarchy of all known creatures. He occupies this distinguished position because he alone is gifted with rational faculties and spiritual aspirations as well as power of action. Man is characterized as God's victory on earth. But the more his rank excels, the more his responsibilities grow. Humans are the dignified beings potentially capable of good and noble achievements. Man has to follow obligations and responsibilities towards God, to our self to other fellow human beings to all other creatures and finally to his physical environment. It is his solemn duty not to waste natural resources not abuse or pollute environment in any manner. It is incumbent upon him to preserve the purity and integrity of the environment so that succeeding generations may enjoy the benefits of God's creations and His blessings.

EXISTENCE OF GOD IN NATURE

Nandita Priyamvada Trivedi
& Sudha Mehta

Environment has two distinct but related parts: the first, which is also, the basic part, is known as our natural environment; the second which is artificial and man made is called our cultural and social environment. The natural environment is obviously more basic. It provides the stimuli for most of our behaviour. It also provides the resources necessary for the satisfaction of our various needs. Not content with the satisfaction of these needs man used his superior intelligence to acquire an understanding of his natural surroundings and has developed the skills necessary for creating his cultural surroundings. In spite of all the development that we have achieved we are still dependent upon our environment. Though this natural environment has been obscured by our cultural surroundings at the same time we use our culture to increase the bounties of nature. Natural environment is the sum total of all the physical objects and the invisible forces and processes in our surroundings which exist and operate entirely independent of human effort (Rollo E. Wicks. Man & Modern Society. p. 20).

We human beings tend to underestimate the vastness, content and complexity of nature. History is proof enough that most of man's achievements and failures can be attributed to his understanding or ignorance of his natural environment. And today as in the past just as man makes demands on nature, nature makes demands on man.

Cultural environment plays a major role in transforming human beings from the man-animal into socialized human beings capable of cooperating with our kind in a more purposeful and effective manner than can any other species.

The term culture is used freely. The common use of this particular word is generally vague. Some people might be using the word culture as the ability to create or appreciate the best in literate or the fine arts. According to another concept culture identifies with the kind of broad, general education to be obtained from a liberal arts course. Culture is also regarded as a possession of those displaying a marked degree of refinement in respect to dress, manners and speech.

Many sociologists and anthropologists define culture as to include man's material creations (civilization) as well as his non-material achievements. Thus culture may be defined as the sum total of the knowledge, beliefs, ideas, ideals skills and ways of acting acquired by men from their group life. Culture is often spoken of as a social heritage, something that is ours without our having created it. Also our culture is social in the sense that it has been produced, preserved and transmitted to us through the co-operative efforts of many individuals. In recent times the concept of culture has been limited to the non-material, such as ideas, beliefs, knowledge, customs and learned skills.

According to a theory of religion it came into existence due to nature's mysteries. Religion forms the base for our cultural environment. But any discussion on religious mostly depends on our willingness and ability to achieve an objective attitude towards it. This is very difficult. We human beings are reluctant to face facts, which seem to conflict with ideas, beliefs and feelings which are as personal and emotional as those pertaining to religion. Ideas and behaviour, which may seem strange to one group of people, may be truly religious to others. And this is exactly how a very personal cultural environment comes into existence. Hence, it is extremely doubtful that many of our own religious beliefs and rituals would make any sense to those who do not understand them. On the one hand the concept of religion ought to be broad enough to include a great range of beliefs, feelings and behaviour. And on the other hand it is extremely difficult to develop a genuine appreciation of the extent to which religion influences human behaviour.

Religion as a concept may be explained as consisting of beliefs, feelings and behaviour. Religious beliefs are convictions concerning the deeper mysteries of life. Beliefs are always learned and uncritically accepted, thus they may be strengthened or altered as a result of experience and reflection. Religious feelings may be defined as emotional states, which are evoked by religious beliefs and convictions. Intensity of religious feelings varies from religion to religion, also from individual to individual. It is not at all necessary that people having the same faith are going to be equally intense in their religious feelings.

The most important factor is religious behaviour. Religion is of importance in any given society and group of people because beliefs and feelings find expression in human behaviour. Behaviour which is motivated by beliefs and feelings about the nature and of powers greater and higher than those of men may be regarded as religious behaviour and this is very personal. In fact religious beliefs, feelings and behaviour are related and interdependent. Cultural anthropologists have pointed out that religious behaviour takes innumerable forms. For example, feasting and fasting, marrying and refusing to marry, saving life and taking life prostitution and cannibalism may be religious behaviour for those regarding them as such.

Thus religious behaviour is directly motivated by religious beliefs and feelings. But on the contrary, religious acts often become so habitual and routine as to have lost most of their rational and emotional significance for the actor. They may become so ritualized and formalised that they have little meaning for those performing them. This being the case, one can hardly expect those of another faith to understand them or see any value in them. Religious behaviour has its ultimate origin in religious beliefs and feelings and in turn it serves to awaken and reinforce these beliefs and feelings. Religious acts help in focusing the attention of the actor on the spiritual, mystical and supernatural. This indicates that religious beliefs, feelings and behaviour are related and interdependent and function in combination.

The cultural theory in relation to the origin of religion specifies that religion is entirely learned rather than biologically acquired. Thus, religion is a human achievement, a body of beliefs, feelings and behaviour devised by man to meet certain needs and transmitted as a part or the social heritage to the members of succeeding generations (Wicks. Man & Modern Society, page 186). The term 'ethnocentrism' is of importance here. It explains the devotion of a people to its own culture, especially to its own ideas of what is right and wrong. We naturally accept the familiar and approve it and we denounce and reject the strange. According to this principle of religious beliefs and practices people view their own views as right and those of others that are different from theirs as wrong.

Every religious group accepts the need for providing religious training for the younger generation. They spend considerable time, effort and wealth to indoctrinate its future members with its own religious beliefs and behaviour. During the process of socialization common religious beliefs and practices serve very strong social bond holding people together.

Hindu religion is not based on any individual ideology; rather it is a way of life. Leading life in accordance with nature, adjusting with nature, comes easily in Hindu religion. The basic philosophy of Panch Tatva - five basic elements is important, *Vayu, Agni, Jal, Prithvi, Akasa.* Thus worship of nature is a part of Hindu way of life. Tribal religions still depend on worshipping nature as their very existence depends on nature. Beliefs and feelings are manifested in our day-to-day behaviour. This behaviour may have its roots in religious beliefs and it may be a most natural way of leading day-to-day life. Hindu religion also keeps its followers close to nature - preserving and nurturing it as a natural behaviour of religious belief, deep-rooted feelings and age-old customs. Hence Hinduism provides its followers their very own cultural environment.

Hinduism is the indigenous religion of India. It has evolved out of the rich culture of ancient India. It has grown over thousands of years into a body of teaching and culture. India is the western name for Hind, the land on the other side of the river Indus. Hinduism is one religion that has no single founder. Early Hinduism developed out of three cultures. Between 3000 and 1500 BC the Aryans migrated to India from present day Iran. In Northern India which was already well established in cities such as Mohanjo Daro in present day Pakistan. In Southern India the Deccan culture flourished for 1500 year up to 1000 B.C. It is not only a religion but also a scientific way of life in harmony with nature. It is the only religion in the world which has not been started by a prophet, but, according to the Puranas, alongwith Brahmanda i.e. nature.

The vedic tradition of Hinduism offers imagery that values the power of natural world. There is an underlying ecological sensitivity within the Hindu tradition. Vedas describe various rituals, which explain in length the earth *(bhu)*, the atmosphere *(bhuvah)*, and sky *(sva)*, water *(ap)* fire and heat *(Agni)* and the wind, *(Vayu)*. Later on in Indian thought these Vedic concepts became formalized in to five great elements: earth *(prithvi)*, water *(jal)*, fire *(Tejas)*, air *(vayu)* and space *(akasa)*.

According to Hinduism the soul can be reincarnated into any form of life. Hence, we find many stories that treat animals as divine. Most Hindus are vegetarians because of their belief in the sanctity of life. Even trees, rivers and mountain are believed to have souls and should be cared for as well as honoured. As Hindus believe in simple living, Hinduism stresses that the source of true happiness is from within and not from outer possession. Life's main purpose is to discover the spiritual nature and the peace and fulfillment it brings. Hindus revere sacred rivers, mountains, forests and animals and love to be close to nature. For example, many villages in India have a sacred lake and around it a grove of trees to catch rainfall and protect the banks from erosion. Thus rainfall is stored to irrigate nearby fields and supply village wells with drinking water. The surrounding area works as a sanctuary for wild life.

Hinduism has long revered the tree. Examples of this are visible in the early seals from the Indus Valley cities (3000 BC) where trees have been depicted as a powerful symbol of abundance. India has a long history of forest protection, from the edicts of Asoka to the individual work of various Rajas, to the modern Chipko movement wherein women surrounded trees with their own bodies to stop the destruction of forest. Even today Banyan tree *(Vat Savitri)*, Anwala *(Anwala Nauami)* Neem and Tulsi are worshipped by Hindus.

Rivers have been and continue to be an integral part of the Hindu religion. Civilizations have developed and flourished near rivers, which is a life-giving source (Amrit) to mankind. More than fifty vedic hymns praise the Saraswati, a river (now dry) associated with the goddess of learning and culture. The river Ganges originated from Siva's head in the Himalayas, gives sustenance to millions of modern Indians.

Hindu religion wants its followers to live a simple life. Running after material pleasures can disturb nature's checks and balances. It teaches not to use anything belonging to nature such as soil, coal or forest at a greater rate than can be replenished. For example, do not destroy birds, fish, earthworm and even bacteria since all these play vital ecological roles. Once they are annihilated we cannot recreate them. Thus we can avoid becoming bankrupt and the life cycle can continue for a long time. According to the Hindu religion, all living beings are sacred; they should be treated with respect and compassion because they are parts of God.

Traditionally the rivers of India have always been considered pure. Hinduism offers a variety of cosmological views. On the one hand Vedas, Upanishads and epics present a style of life in tune with the elements. Some examples are given. These lines from Bhagvad Gita indicate that a life without contribution towards the preservation of ecology is a life of sin and a life without specific purpose or use.

Living bodies subsist on food grains, which are produced from rains. Rains are produced from performance of yajna (sacrifice) and yajna is born of prescribed duties. Regulated activities are prescribed in the Vedas and the Vedas are directly manifested from the Supreme Personality of Godhead. Consequently the all- pervading Transcendence is eternally situated in acts of sacrifice. My dear Arjuna, one who does not follow in human life the cycle of sacrifice thus established by the Vedas certainly lives a life full of sin. Living only for the satisfaction of the senses, such a person lives in vain (Bhagvad Gita).

Life is sustained by different kinds of food: rainfall produces food: timely movement of clouds brings rains: to get the clouds moving on time yajna, religious sacrifice, helps: yajna is performed through rituals: those actions that produce rituals belong only to God: God is revealed by the Vedas: the Vedas are preserved by the human mind: and the human mind is nourished by food. This is the cycle that helps the existence of all forms of life on this globe.

One who does not contribute to the maintenance of this cycle is considered as a destroyer of all life here. When the Lord desired to create life, He created the Sun, Moon and Earth and through them a congenial atmosphere for life came into being. Therefore the Sun, Moon, Earth, Stars and all objects in the universe jointly not individually, create the atmosphere for the creation, sustenance or destruction of everything in the universe. The Earth is the only daughter of the Sun to produce children. The Moon is essential for the creation of the right atmosphere for those children to exist and evolve. This we say because of the influence of the Moon on high and low tides in our rivers and oceans. This is narrated also in the Bhagavad Gita. *I become the moon and thereby supply the juice of life to all vegetables.*

We cannot refute this influence of the Moon on life. It is proved by the movement of all liquids on this globe depending on the movement of the Moon. Therefore, ecology in totality must be prescribed: just a part of it would not suffice. Hinduism is a religion that is very near to nature. It asks its followers to see God in every object in the Universe. Worship of God in Air, Water, Fire, Sun, Moon, Stars and Earth is specially recommended. Earth is worshipped as the spouse of God, hence regarded as very dear and near to God. All lives on Earth are considered as children of God and Earth.

Sri Krishna in the Bhagvad Gita says: "I pervade the Universe. All objects in the Universe rest on me as pearls on the thread of a garland." The Upanishads narrate that after creating the Universe, the Creator entered into each and every object to help them maintain their interrelationship. The Upanishad says, *"tat sristva to pravisat"*: ("after creating the universe he entered into every object created"). Therefore, to contribute towards the maintenance of this interrelationship becomes worship of God. Hindus believe that there is a soul in all plants and animals. One has to do penance even for killing plants and animals for food. This daily penance is called visva deva. Visva deva is nothing but an offering of prepared food to the Creator, asking his pardon.

The Hindu religion gives great importance to protecting cattle. In old times, in every house there used to be a cow. It is a great friend of humans. It nourishes us through milk and provides manure to grow our food. It lives on the fodder we get while growing our food. Advanced countries have started to realise the harmful effects of consuming food grown with chemical manure. When we use chemical manure, the topsoil loses its fertility. This generation has no right to use up all the fertility of the soil and leave behind an unproductive land for future generations.

"One who plants one peepal, one neem, ten flowering plants or creepers two pomegranates, two oranges and five mangoes do not go to hell". (Varsha Purana). "Five sorts of kindness are the daily sacrifice of the trees, to families they give fuel, to passers by they give shade and a resting place, to birds they give shelter, with their leaves roots and bark they give medicines." (Varsha Purana) "The rivers are the veins of God, the ocean is his blood and the trees the hairs of his body. The air is His breath, the earth His flesh, the sky is His abdomen, the hills and mountains the stacks of His bones and the passing ages are His movements." *(Srimad Bhagwad Gita).*

"Oh wicked person, if you roast a bird, then your bathing in the sacred rivers, pilgrimage worship and yagnas are useless." *(Srimad Bhagwad Gita).*In India people belonging to Hindu religion have a wonderful way of welcoming and celebrating change of each season with certain religious festivals and rituals. These are ingrained as our cultural heritage. In many other countries various seasons are welcomed in somewhat similar fashion. These ages old traditions are followed even today by the young Hindu generation. They may not have a liking about the origin and reason for celebrating these festivals. Some examples may be given, such as Makar Sankranti (Pongal Lodhi), which implies the change of Sun. They also indicate the maturing grains. Similarly, *Shard Purnima* indicates the arrival of winter, etc.

There is no life that is inferior. All lives enjoy the same importance in the Universe and all play their fixed roles. They are to function together and no link in the chain is to be lost. If some link were lost, the whole ecological balance would be disturbed. All kinds of life – insects, birds and animals-contribute towards the maintenance of ecological balance, but what is humanity's contribution toward this? Humans are intelligent animals; therefore our contribution should be the biggest. But it is not. We are nullifying the benefits of the contributions made by nature. Traditional Hindu practices of caring for nature are being forgotten and as a result human survival is becoming more difficult.

It may be concluded that feeling one with nature is the fundamental environmental message of Hindu culture. Hindus perceive life not only in human beings, but also in plants, birds and animals. This vision of oneness of life has helped Hindus to develop a worshipful altitude towards everything in nature. Nature is not a commodity to be dominated and conquered. Man must change the attitude of dominating nature to one of co-operating with it. A fundamental reorientation of human consciousness is required to recognize that earth has the same relationship with man as mother with her child. Life is an organic entity and the sea, soil, mountains plants and animals are inseparable parts of the cosmic web. Man must learn to live in harmony with nature. Man must recognize that plants and animals have a meaningful life too, in the cosmic play (Lila). Environmental issues require a spiritual response. An awareness of the ecological balance ought to be created at all levels of human thought and activity.

SYMBIOSIS-FLORA, FAUNA AND HUMAN LIFE

Nirmala P.

"And God blessed them and said unto them, "Be fruitful and multiply and replenish the earth and subdue it and have dominion over the fish of the sea and over the fowl of the air and over every living thing that moveth upon the earth." That is Genesis One, verse 28 and the Bible is a chronicle of the beginnings of our civilization.

In the modern context it comes as a mighty disappointment after verses 1-25, in which God was jubilantly summoning up light and darkness, water and firmament and all other magnificence of raw nature and at every stage stepping back and saying, "this is good." By the end of verse 28 God's interest is only on humans.

The Western world's attitudes towards nature were shaped by the Judeo-Christian tradition, which involved the concept of a world, created solely for the benefit of man. In older religious traditions, humans were seen as part of nature, rather than the rulers of nature. In animistic religions, there was believed to be a spirit in every tree, mountain or spring, and all had to be respected. In contrast with paganism and Eastern religions, Christianity not only established a dualism of man and nature but also insisted that it is God's will that man subdue nature for his proper end.

A liberated view of religion suggests that we are stewards of God's earth, caretakers ruling over that which is not ours. God desires that we treat the creation in the same way as God treats us. God loves the world, of course not in its sinful strivings and unholy notions. God loves the world for the sake of its origin; because God has thought it out; because God has created it; because God has maintained it and maintains it to this day.

These conservation and humane precepts of the Scriptures were well understood by the early leaders of Judaism and the Christian Church. Many early Christian saints are depicted as having close and friendly relationships with wild and domestic creatures. St. Francis spoke to birds, took to lepers and to the wolf of Gubbio. Theologically, as God's aim is to evoke God values enjoyments that at all levels, it means that reverence for our neighbours extends to reverence for all creatures. "God is love", is an important assertion in the process theology. All of this thinking is based on the firm conviction that minerals and plants are alive and that their magnetic fields can be variously influenced to the benefit or disadvantage of mankind. Now, looking back over time, at man's departure from the eco-stewardship to which God called us, we can see that our eco-abuse has taken its toll. Man has sowed a crop of trouble for the future. And the future has arrived!

Creation is being degraded through human abuse, indifference, and negligence. We often are on the sidelines watching its dismemberment if not participating in its destruction; permitting many species to be extinguished, not speaking out against rampant soil erosion, obscuring the testimony of the heavens with our polluting wastefulness, becoming the only occupants of the land, muddying the waters, trampling the vegetation, and much more. The world God loves is being subjected to destructive use even while we debate how God made it. As we (seemingly) defend Creation by our words, we may find that we destroy it by our deeds of omission and commission.

Among the many areas of prime concern today, is the state of our environment. The environment is an inseparable whole, constituted by the complex interacting system of physical, biological and cultural elements, temperature, humidity, atmospheric pressure and sunshine-all interlinked individually as well as collectively in myriad ways. Our present plight is the widening gap between environmental change and environmental understanding. The eco system includes living organisms and non-living environment, both necessary for maintenance of life. Everything in the Universe is related to and affects everything else. This is the principle of the environmental unity and is termed as symbiosis.

Humans, plants and animals have existed together for many years without creating loss of species. Each species has a role to play and importance of its own. Throughout the ages it has been widely believed that, mankind could enjoy health and happiness by living in accordance with the ways of nature. But man abandoned the ways of nature, in the very process of becoming human.

But the case of man is curious in that, the population of human beings is rising by leaps and bounds and at the same time, man is exploiting natural resources much faster than these can regenerate and replenish themselves. Man's first act of civilization was destruction of forests for human dwellings and for cultivation. Symbiosis is a state in which, members of different species habitually live together in physical contact, mutually benefiting, from each other's presence. Most of these interactions involve food. Some of the interactions are often brief while some extend to longer periods. The lives of the symbiote and symbiont are inextricably bound together so that the species cannot live apart from each other.

Green plants use carbon dioxide for photosynthesis and in return release oxygen to the atmosphere. Animals inhale oxygen during respiration and give out carbon dioxide. Thus a continuous set of actions and interactions take place among the living organisms of the biosphere. The one sets a balance in the environment and the other upsets the balance.

Animals are incapable of transforming sunlight, metabolically into usable energy and so must eat plants directly or be carnivores. Plants, therefore, become the first kingdom of self-reproducing creatures. To accomplish this adjustment, many other factors become involved, even the animal kingdom must participate.

Plants are producers. They synthesize their food from inorganic and organic materials of the environment. Consumers consume the food produced by the producers. The rabbit forms the food of wolves, which are primary carnivores. The wolves are taken as food by the lions, which are secondary carnivores. This is known as food chain. Decomposers, which include bacteria and other microorganisms, decompose the dead bodies of producers and consumers and reconvert them to their original components-water, carbon dioxide and mineral salts thus balancing the ecosystem.

Animals use plants as food or for transport, shelter, substratum or dispersal of seeds. An American robin benefits by building its nest in a red maple tree. The shell inhabited by the hermit crab is the abandoned house of a snail. Sometimes the base of the anemone grows over the entire shell, and or the shell is dissolved away so that, the crab's house consists almost entirely of the anemone's base. The sting of the anemone protects the crab. The crab keeps the shell clean and the anemone in return snatches fragments of food from the pincers of the crab.

Birds, mammals and other animals eat fruits. The seeds are dropped in the excrement at varying distances from the original site. The pollination of flowers is another manifestation of symbiosis in which the species concerned may be in contact for only a few seconds. The insect gets nectar and in return carries pollen for ensuring cross-pollination. Sometimes the adaptations of the flowers and insects are often so advanced that only a specific insect can successfully pollinate a flower.

The oxpeckers land on grazing herbivores and eat away ticks and other parasites that live on their skin. The oxpecker gets food and the herbivore gets pest control. Besides, when there is danger the oxpecker flies upwards and screams a warning, which helps the symbiont. Another kind of pest control service is rendered by the crocodile bird in removing leeches from around the teeth of the crocodile, which allows the bird to enter the mouth for the search.

A certain kind of bacteria lives in the intestines of human and many other animals. The bacteria partially digest food that the humans cannot digest allowing the human to finish the job. Earthworm, by churning up the soil empowers the growth of plants. Some of the bacteria in our large intestine supply us with vitamin K.

Predation is the law of nature. It lays the foundation of ecosystem structure and function. It is wrong to describe the hunter as exploiting the prey. The prey represents an expendable link in the trophic chain. Though generally the hunted species is more than predator species, any abnormal increase in prey population automatically leads to unusual increase in the abundance of the predator. This is followed by abnormal hunting activity, so that the prey population becomes decimated and with it the predator population also declines. Predatism in course of time passes off into cannibalism, parasitism, carrion feeding and scavenging. Carrion feeders and scavengers maintain general sanitation of habitats.

The deer tick gets food and a warm home from the deer without killing it. Losing blood and possibly getting an infected wound harm the deer. Hosts also develop ways of getting rid of the parasites. They can scratch away ticks. Some hosts build a symbiotic relationship with another organism that helps to get rid of the parasite. Ladybugs live on plants, eating the aphids and benefiting by getting food, while the plant benefits by being rid of the aphids.

"Industrialize or perish" has been the watchword of the eco development. Paradoxically industrialization turned out to be one of the most threatening factors contributing to environmental imbalance. Progress of globalization in the third world countries further escalates environmental tragedies.

Forests are central to civilization, culture and faiths. Deforestation is identified as one of the serious ecological catastrophes of today. Forests are cleared for construction of roads, residential quarters, storage of building materials, etc. Destruction of extraordinarily rich nature and the loss of precious flora and fauna, which form the main source of life and fertility, lead to soil erosion, floods, and destruction of life, water logging and salinity. Illegal timber felling is the surest and easiest way to become rich. To increase revenue, extensive mixed forests were converted to mono cultivation. This led to loss of habitat to thousands of life species like birds and insects. They also absorb ground water and eventually dry up the soil.

In India, The Himalayan region is full of high mountains, rich vegetation, flowers, pastures and glaciers. It keeps the balance of atmospheric circulation over much of the Indian continent. It gives rise to river systems with tributaries, which pass through the mountain valleys covered with forests. In the plains they become source of life for millions of people.

Destruction of forests affects radically the very life and culture of the tribals, and the poor forest dwellers, as their life styles are inextricably interwoven with the forests. Landless labourers, marginal farmers, sharecroppers, and those who live by collecting and selling forest products are forced to pay a heavy price. Small farmers who supplement their meager income by ranching cattle are hard hit as they are bereft of enough grazing land in their new settlements. Forest dwellers depend on forests for food, fodder, fertilizer, construction material, medicine, etc. Herbal and traditional medicines and edible plants are no more readily available. Women are forced to walk long distances to fetch water and collect fuel. With scarcity of food, less food is given to female members of the family to ensure that boys and men are fed relatively adequate. Men uprooted from their homeland, culture and faith are forced to migrate to urban areas in search of jobs. They tend to resort to violence, rampant poaching or have become bonded labourers or urban migrants who end up as the men of the street. Deforestation alongside rapid industrialization causes melting of ice at the poles leading to rise in sea levels and low lying countries will be worst affected. Underground water pockets disappear and soil loses its nutrients. Rare wild life, e.g hoolock apes and hornbill are extinct.

Dam construction activities have led to irreversible fragmentation and quantitative degradation of the remaining vegetation, which also affects flow of rivers. Clearing of forests, for dam projects ended in cutting trees in unaffected areas too. Utilization of water resources through large dams has resulted in excessive submersion of valley forests in the river catchments.

Air pollutants emitted from mines, power plants, industry and motorized transport contain Sulphur dioxide which mixed with water in the clouds falls as acid rain or snow or mist. Acid rain, depletion of ozone layer, and green house effect increase the heat of the atmosphere. About 95 % of the sun's ultra violet rays are absorbed by a thin layer of ozone which lies in the stratosphere. It protects the Earth from exposure to damaging levels of radiation. This results in skin cancers and other diseases too.

Modern civilization has increasingly eroded the symbiotic bonds between humankind and nature. No doubt the twentieth century has been a catastrophe from the point of view of environment. The story is repeated on the coasts and beaches as builders destroy mangroves, kill marine life, salinise and coastal aquifers to build hotels and prawn farms. To the other side there are millions cocooned in the air-conditioned homes, offices and cars, looking at nature through the magnifying glasses of posh hotels in unpolluted hill resorts. With more unrest brewing, the number of environmental refugees is growing as we propagate Coke, Pepsi and McDonald's.

Animals and plants have succumbed in far greater numbers, helpless against competition from the most rapacious of all Earth's animals. Towns have expanded eating up woods and grasslands. Bulldozers have obliterated great tracts of countryside. Mines, mills and factories have poisoned rivers. We have cut down forests for timber, drained marshlands and dammed great valleys. To feed our growing numbers we have deprived wild animals of food. We have let domestic animals overgraze pastures. We have allowed rivers to silt up, often rendering them lifeless. We have killed animals for food and shot them for sport. We have over fished the rivers and still are doing so. Our motor vehicles kill millions of animals every year. Wherever our ancestors went they took plants and animals often with devastating effects. They introduced species that have run wild in many parts of the world, sometimes wiping out native creatures. We have put brown and rainbow trout, to satisfy the passions of anglers. As the world's wild life vanishes, demand rises among collectors and tourists for rare specimens. Poachers raid sanctuaries for skin ivory and horns. Others smuggle live animals, bird and rare plants across the world's frontiers. Some 40% of the drugs essential for our health come from plants, microbes and animals. Yet scientists have examined only three per cent of the world's flowering plants for possible drugs. Only by working with nature can humanity survive.

Nature once frightened us and made us tremble with its mysterious vastness and uncontrollable energy of its elemental forces. Now it frightens us with its limitations and newfound fragility and plastic mechanisms. The human feedbacks into Nature have become increasingly disharmonized. Man's means of rational survival in the environment now damages the biosphere on an increasing scale and on the boomerang principle affects man himself. Man's protracted interference with the balance of Nature has at last opened his eyes to a fact long ignored organisms of the earth don't exist independently or uninfluenced by their surroundings. Man is becoming increasingly aware that he lives in a world filled with life rather than mechanistic processes. He is beginning to estimate the importance of maintaining natural order and recognizing that all forms of life are indispensable to the common survival of the universal plan.

As humans we should protect our flora and fauna, realize their value, appreciate them, and ensure that they have places to live in the wild. But not the least, we should silently thank them for all they contribute to our lives. They truly enrich and nourish our human experience. To resolve these problems we must build a comfortable and healthy living environment where humans and natural environment coexist, where the environment is managed in a sustainable fashion and where nature's stern discipline enjoins mutual help.

We should "Think Globally and act locally". Drastic changes are needed. It is time to courageously think globally breaking away from traditional paradigms and plunge boldly for a better environment.

Scientific breakthroughs and technological ingenuity have given us the capacity to overcome all the environmental challenges. What is missing is the wisdom and compassion to do so. It is this yawning chasm that will need to be bridged if we are ever to reverse the present trend towards disaster. Environmentalists and activists should create public awareness and foster intelligent debates. A new culture of convergence and cooperation must emerge if the rich promise of the next millennium is not to dissolve into conflict and chaos.

Karl Marx advocates Humanization of nature and naturalization of man. The world of nature expresses the freedom and delight of God in creating. The creative breath of God is inspired by beauty and joy. The most powerful energy to transform our world, the energy most needed today is the spirit of love. Love alone is capable of creating one earth, one human community; love is the spirit of one earth, the life giving transforming dynamic at heart of the noosphere.

God is present in creation. God enjoys and shares our sufferings. God appears in every being, beckons in every relationship and bursts forth in every ecosystem. The account of Genesis1:1-24 "The earth is the Lord's, and everything in it, the world, and all who live in it." Psalm 24 offers that God's sovereignty and lordship is over creation and not human dominion. It follows that the human beings' responsibility towards God's creation.

Man would fail in his royal office of dominion over the earth if he were to exploit the earth's resources to the detriment of the land, plant life, animals, rivers and seas.

God expects mankind to be Viceroys to his creation and the Viceroy on earth is expected to defend the cause of the poor and be concerned about the fertility of their land. Christ is the principle of cosmic cohesion making creation a cosmos instead of chaos. Man's unity with nature is part of the created world so he should use it for God's glories and not to exploit it. In God's peaceful kingdom:

> The wolf shall live with the lamb,
> The leopard will lie down with the goat,
> The calf and the lion and the yearling together;
> And a little child will lead them.
> The cow will feed with the bear,
> Their young will lie down together.

Bibliography

Odum, E.P. (1971), *Fundamentals of Ecology"*, W.B. Saunders Company.

Agarwal, A. and Narain Sunita (1985), "The India's Environment, the Second Citizen's Report", edited, Centre for Science and Environment, New Delhi.

Gadgil, Madhav (1989), Deforestation, Problems and Prospects, SPWD, New Delhi.

Roy, Sunil, K. (1933), "Drowned in the Deluge of Development", The Times of India Aug 13.

EARTH, NATURE AND LIFE: A MEETING POINT

Shobha Diwakar & Neelanjana Pathak

"When man does not realise his kinship with the world, he lives in a prison-house whose walls are alien to him" (Tagore, *"Sadhana")*. The history of man from the beginning right from the travelogues of Vasco de Gama, Ibn Hakul Masudi, Herodotus- expresses man's innate curiosity to delve deep into the mysteries of the world that he inhabits. Nature, Life and Earth have been engaging themes from earliest times as man has tried to scrutinize his affiliations with the three and also look for their meeting point. The three have had a crucial role in all stages of human evolution irrespective of any cultural, linguistic, or geographical differences. Life, Nature and Earth at first sight may appear to be apart from each other, but at deeper level of understanding, the three are a part of each other. They are complementary aspects of the Absolute that co-exist without any conflict, making this cosmos one composite whole. This paper is an attempt to show that the apparent divisions are limited to the gross aspect, but at the subtle level, the ground, view is of essential oneness. An endeavour will be made to certify that in identifying one transcendental identity in the three and in seeing a common spirit in animate, inanimate, conscious or unconscious aspects of Nature, lies the solution to the grim ecological imbalance that stares us today. In threading the three together we shall establish familiar ties with all humans and with the animal kingdom, the flora and the fauna.

Richard Park, An American sociologist, observed that "human ecology deals with the web of life, the balance of nature, the concepts of competition, dominance and succession, biological economics and symbiosis (Dixit 234). When we talk about Earth, Nature and Life, we refer to man's spread over the earth and his connections and interactions with the manifestations of nature. Nature is an authentic sign of God's presence; since it proclaims the glory of God; it is only natural that spiritual and religious connotations began to be sought in it. It has been an accepted opinion from the earliest tradition of the protoplast that there is one Absolute Power-the maker and ruler of heaven, earth, nature and all living beings. Consequently, this cosmos reveals its Author, Designer or Artist often addressed as God. From pre-Socratic times, Greek thinkers had shown intense interest in seeking the original principle from which all things had issued. From Thales to Anaxagoras, Anaximander and Anaximenes- all believed it to be either water, air or some other form of nature. To the Greeks, Nature is a goddess, a God infused entity, a spirit, a state of flux, a half-true myth, a game of atoms, a signature of God's oneness. In the *Rigveda* Nature is understood as a cosmos governed by a law of truth, executed by God, sharing man's joys and sorrows. The mysterious workings of Nature are unfathomable by the human mind.

Mother earth or *Dharti Mata* for Indians, is a mother because it has in its womb the seeds of man's sustenance and nourishes every life. It bears upon itself the weight of all Nature and Life. Mortals harness it for agriculture and dig its womb for the treasures that lie within its layers. In XII. I of the *Atharva Veda* the Earth is called the perfume of creation, the aroma of God's energy. It is also the summation of all the other elements and all qualities of matter:

> May that Earth, which holds on Her ample lap the ocean,
> The rivers, the lakes, which bears the crop yielding soil,
> And is ground for all that breathes and stirs and lives,
> May she also bear the fruits of what we have done.

According to *Encyclopedia Britannica,* Life is "State characterised by the ability to . . . grow, reproduce and respond and adapt to environmental stimuli". (64). Life means different things to different people so generalisations may be misleading. However, it is certainly not merely an evolutionary process of well-organised molecular reactions. The question of life is related to philosophy, theology and metaphysics. To the spiritualists, life is a divine spark. The Epicurean view of the origin of life, of man, of society is:

> First the earth grew vegetation, much as animals grew hair and heat and moisture generated animals. Then she brought forth man from innumerable wombs of her contriving. In the beginning man led a purely animal existence, completely at the mercy of natural forces. (qtd in Fuller 140)

Just as in nature, there is a continuous struggle for existence among various species of plants and animals, so is there within the human world because life depends on the earth's fertility and nature's productivity. Nature's amiability and benevolence foster healthy living conditions as seen in the Indian peninsula or almost all the wet tropical and subtropical regions. Where there is 'God's plenty' as in South and East Asia, Nature is revered. However, the Hebrews and Greeks struggled for long against the hardships of unrelenting natural conditions. They had to learn to vanquish Nature in order to evolve and progress. Consequently, in the European cultural realms; the environment came to be seen as a metaphor for triumph over struggle, for dominant forms of social management over the weak. . ." (Dixit 242).

There are two easily recognizable trends with respect to man's ties with Nature. On the one hand, there are the conservatives to whom reverence of Nature is equated with morality and virtue; on the other there are the rationalists who are driven by the urge to compete in materialistic terms even if it means sacrificing ethics. For the conservatives, there is an inextricable tie between Nature, Life, Earth and God. They are convinced that God first created the Earth and then created man to respect the bounties of nature with a sense of awe and wonder. As Thomas Pope Blount put it poetically:

> Every flower of the field, every
> fibre 'of a plant, every particle of
> an insect, carried with it the
> impress of its Maker, and can-if duly considered-read us
> lectures, on ethics or divinity. *(World Poetry* 1026)

This school believes it to be a moral obligation to preserve and care for the environment. With the rationalists or modernists, the order altered to God--Man-Nature. They believe that nature needs to be harnessed and even exploited if need be to serve human ends. The view of the conservatives is identified by a critic as close to Garianism-a term derived from Gaia the Greek goddess, the nurturing mother figure. Interestingly, Garia's daughter Themis was the goddess of justice. Thus, justice was equated with retribution-those who disturbed the environment would be punished; those who respected it would be rewarded:

> The Earth has her own law, a natural law. . . deeper than human enactments and beyond repeal. . . Earth forgives but only to a certain point, only until the balance tips and then it is too late (Dixit 248).

Indian life and thought have primarily been governed by ecological consideration Plants, rivers, trees, reptiles and some animals are considered sacred. This augurs compassion for all forms of life, human and non-human. Planting trees in the courtyard and quenching the thirst of weary travelers are pious duties directed to preserve life and environment. Manifestations of Nature have a crucial role in shaping attitudes-it is from Nature that we learn the first lessons of adaptability according *to* climate, season, and soil conditions. Ellen Churchill Semple writes:

> Man is the product of the earth's surface. This means not merely that he is a child of the earth, dust of her dust, but that the earth has mothered him, fed him, set him to task, directed his thoughts;... (qtd in Dixit 105)

Indian thought has been both religion and nature based because of the emphasis on practical realization of truth. The word *darshan* means "Vision" and stands for immediate and intuitive vision of that reality which leads to realization. The most important theme of the *Upanishads* is to know Nature and establish a kinship with it. In both the Indian epics the *Ramayana* and the *Mahabharat* there are representations of the horrific and thinking aspects of Nature and at place there are direct parallelisms between human deeds and natural events. The Vedas are as old as Nature because they represent the divine voice, which emerged at the time of Creation of Nature. Amazingly, even the ozone layer is mentioned in the hymns of *Rigveda* and *Atharva Veda* as a thick membrane surrounding the earth:

> *Tasyot jayamanasya-ulba a sida hiranyaya.-Atharvaveda* 4.2.8. '*Ulba*' is the membrane while the world '*hiranyaya*' describes the golden colour. The fundamental idea of oneness of spirit is also expressed repeatedly in 'the Vedic hymns that indicate that one common power of creation, with the same origin animates rivers, mountains, winds, the flora and the fauna. According to Indian philosophy, all living organisms return to the *Panch Tatva* or the five elements, which consequently may be taken as a meeting point of Life, Earth and Nature.

The African tribes and the native Indian Americans have been known to have worshipped different forms of nature. They have been struck by its edifying and compelling energy and saw in it some supernatural powers beyond their control. China, the oriental Buddhist nation remained untouched by the wave of scientific enquiries and continued to believe until the last century what saint Laa-Tse had preached centuries ago; The world moved according to a divine pattern, reflected in the rhythmic and orderly movements of nature. The sum of wisdom and of happiness for man is that he adjusts himself to this order and he reflects the way the world moves. (Ross 76). Thus, Tao philosophy again revolves around the interdependence and interaction between man, nature and earth. It is their conviction that "the man who tries to shape the world into what he wants it to be, damages himself and others in the attempt" (Ross 75). Even in Japanese *Shintoism* there is a deep spiritual kinship with Nature. They also believe that the same wonderful forces that move in nature move in themselves. There is no dividing line between Life, Earth and Nature. Just as the Chinese and Japanese worship Nature, the Jews celebrate the renewed fertility of the earth' during springtime. To them, spring is a metaphor for celebration and renewal of nature and all forms of life. The Jews strongly believed that the Universe is governed by law, order and purpose.

Thus, for thousands of years in all ancient cultures, nations and tribes Earth, Life, Nature and Man were regarded as accruing from a common source: A Supreme Power, omnipotent and dextrous beyond the periphery of human imagination. This resulted in awe, which changed to fear and finally to reverence. With the passage of time, man's knowledge of the world around him has widened. Scientific enquiry has deepened and new ideas are emerging as science and technology is altering man's perceptions. A scientific world based on reasoning is replacing the world of faith. Theories regarding Nature are being re-examined and re-written. The awe remains. The fear is overcome. The reverence must renew.

The dark ages, the Time Cycles of almost total, abysmal ignorance are left behind us and new vistas are before us as far as we are willing to view them with our inner vision, if we adequately awaken to a sufficient degree of realization of our essential beingness. The panacea lies in seeing the spark of divinity that threads Earth, Nature and Life. As sentient beings, we have to foster a harmony with natural laws, which are spiritual consciousness .Only narrow and fragmentary reactions to Life, Nature and Earth may make them appear apart from each other, but a spiritually enlightened being, recognizes the underlying identity of the provisionally differentiated entities-Nature, Life, Earth and also man. Even a blade of grass can teach us lessons of Oneness, of identity with the greatest and smallest manifestations of Nature. American poet Walt Whitman believes that the grass which grows over a grave illustrates the idea that Earth, Nature and Life renew themselves and follow a cyclic pattern for regeneration:

> The smallest sprout shows there is really no death,
> And if ever there was it led forward life,
> and does not wait at the end to arrest it ("Song of Myself 06").

The unity of life in nature upon this earth can be re-affirmed even by some scientific discoveries. J.C. Bose proved by research that there is a detailed resemblance between the circulatory systems in plants and man, and that the plants when subjected to stimulant depressants or poison would cut short their life. (Durant 46)

It is ironical that the once glorified earth is today a blanket of dust, chemicals, industrial and vehicular pollution which have caused depletion in the protective ozone layer above the earth's atmosphere. Dangerous radiations from the space and the rising temperature of the globe have resulted in an alarming rate of killer diseases. Are these the boons of science or are they the curse of moving away from Nature? The past few decades have witnessed several attempts- environmental protection projects, policies, protocols- at the local, national and global levels. Yet, the problems shall remain till we introspect to examine where and how we have failed in preserving our kinship with the Earth we inhabit, the Nature that sustains us, and the Life that surrounds us.

In order to strike a balance between the inward and the outward, life and nature must be ecologically balanced. When *Shakti* and *Karma* are interwoven, a moral order is established. This is what the ancient texts have taught us and we need to remember it even if we have taken giant strides towards scientific innovations. Huxley calls man the "trustee of the evolutionary process on earth" (135). Man's irresistible hunger to bring Nature under his command has disturbed the ecological balance and has proved his failure as a "trustee". His entry into the forbidden territories of the cosmos could be taken as a final assault on Nature. The peace of the moon, a natural satellite to scientists, but Diana the Greek Goddess, to believers has been ruffled by the landing of spacecrafts. The discovery of the surface being made up of only rocks and Rocky Mountains shattered all myths and legends that had lent it an aura of mystery. Such scientific discoveries disillusion man and may widen the gap between science and religion. Gradually, man will lose the fear of the sinister power of a privileged hierarchy (Huxely 136). Man cannot live outside the web of

Nature-nor can we escape the crucial influence of Nature because, " . . . every aspect of man's life upon the earth, (including his philosophy, religion),. . . mirrors the influence of the physical environment of his habitat" (Dixit 105). Certainly, life means more than food and the body, more than shelter and clothing-it is the dexterity to reconcile our inner self with Nature and Earth and fathom the essential Life flowing through them. In this lie the triumph of Man's life and the symbiosis of Nature, Life and Earth.

It is paradoxical that man, the intelligent being has built the grand edifice of progressive civilisation and scientific evolution with near total neglect of the foundational support system of Nature and Earth. He needs to realise that the quality of life on earth is directly proportional to the responsibility he shows towards Earth, his habitat, and Nature his life supporter. Scientists may hype the benefits of molecular biotechnology, gene-cloning, human genome and other laboratory techniques using inanimate matter, but the fact remains that these cannot create living cells. Man has walked on the moon, made successful forays into outer space, and achieved material prosperity, but has failed miserably in establishing equilibrium with Nature's law. Outward progress has become synonymous with inner turmoil, which cannot be cured by spread of religion, large congregations, ritualistic observances and discourses of saints. The need is to fine-tune the inner conscience with the one all pervading spirit wherein meet Earth, Life and Nature. Nature-gentle, calm and tranquil serves as an escape, as a counterpoise for man's hectic life in courts and cities. It's permanence serves as a contrast to the temporariness of life.

Whatever Nature reveals about itself through the sense and beyond perceptions to the human beings is what they understand it to be. Species other than human-beings are not mentally evolved to conceptualize that the movements in the universe and in them are controlled by the one Omnipotent, Omnipresent, Absolute Power as their soul force. The excellence and superiority of the homo-sapiens lies herein-it is they who can conceptualize Nature, Earth and Life in their myriad forms and discover a divine spirit in which lies their meeting point.

May this become for us a solemn occasion to rededicate ourselves to resolutely work towards the betterment of Earth, Nature and Life. With the consciousness of one divine spirit, the image of God Himself in Nature Earth and Life "Nature and man shall be disjoined and diffused no more" (Tagore "Passage to India" 149). When the realization of an essential affinity in all manifestations of God, dawns upon the sceptical human mind, the ideals of *Atmavat Sarvabhooteshu* (all living beings are soulkins) and *Vasudheva Kutumbkam* (Entire Earth is our family), will narrow the distance between man and man, man and Creator, man and Nature, man and Earth, man and Life-this will be the meeting point of Earth, Nature and Life.

References

Dixit, R.D. (2001). *Geographical Thought: A Contextual History of Ideas.* New Delhi: Prentice Hall.

Durant, Will (1966). *The Pleasures of Philosophy.* New York: Simon and Schuster.

Fuller, B.A.G. (1987), *A History of Philosophy.* New Delhi: Mohan Primlanu.

Huxley, Julian (1953), *Man in the Modern World: A Mentor Book.* New York: New American Library.

"Life". *Britannica Ready Reference Encyclopedia* Vo1.6. New Delhi: Encyclopaedia Britannica Pvt. Ltd., 2005.

Ross, H. Floyd (1956), *The Great Religions by Which Men Live, A Premier Book.* Fawcatt World Libra.

Spiller, T.S. ed. (1969), *Anthology of World Poetry.* New York: STace Publishers.

Tagore, Rabindranath (1988), *Collected Poems* Ed. S. Roy. New Delhi: Harper Collins Publishers Pvt. Ltd.

Vastyayan, V. (1983), "The Atharava Veda XII. 1". *The Indian Poetic Tradition: An anthology of Poetry.* Agra: YK. Publication.

SYMBIOSIS-FLORA, FAUNA AND HUMAN LIFE

Jacintha & Rama Devi

The earth is a beautiful and fascinating planet of the universe. It has provided optimum conditions for development of life. The resources and environment of the planet have provided us great heights in arts, science, technology and other pursuits. In nature the flora and fauna live independently, co-exist and interact with each other. The inter-acting, inter-relating flora and fauna coupled with natural resources constitute the environment. The flora of the planet by the process of photosynthesis, supply the O^2 for the respiratory activity of living organisms on the earth. It supplies food to the living, and also cleaning the atmosphere by consuming the CO^2 of the atmosphere, hence called "the lungs of the atmosphere".

We have thousands of edible plants and animals over the world, out of which only about three dozen types constitute the major food for human. The main food reserves include wheat, rice, maize, potato, barley, oats, cassava, sweet potato, sugar cane, pulses, sorghum, millet, about 20 or so common fruits and vegetables, milk, meat, fish and seafood. Amongst these, rice, wheat and maize are the major grains, about 1500 million metric tons of which are grown each year, which is about half of all the agricultural crops. About 4 billion people in the developing countries have wheat and rice as their staple food.

Meat and milk are mainly consumed by more developed nations of North America, Europe & Japan who consume about 80% of thc total. Fish and sea-food contribute about 70 million metric tons of high quality protein to the world's diet. But there are indications that we have already surpassed sustainable harvest offish from most of the world's oceans. A large number of wild animals is also our source of food. The plants are being exploited as food sources, fruits and vegetables, but also utilized for various other needs by man like spices, condiments, other flavourings, fumitory and masticatory materials, wood, vegetable tannins, dye stuffs, rubber, insecticides, herbicides, drugs and medicines. A large number of marine animals is supposed to possess anti-cancer properties which are yet to be explored systematically. The fossil fuels, coal, petroleum and natural gas are also products of fossilized biodiversity.

The animal products like tusks of elephants, musk from musk deer, silk from silk worm, wool from sheep, fur of many animals, lac from lac insects, honey, etc. all of which are traded in the market. Many industries are dependent upon the productive use values of biodiversity like paper and pulp industry, plywood industry, Railway sleeper industry, silk industry, textile industry, ivory-works, leather industry and pearl industry. Wild life products are smuggled and marketed in large quantities to some rich western countries and also to China and Hong Kong where export of cat skins and snake skins fetches a booming business.

The values associated with the social life, customs, religion & psycho spiritual aspects of the people are social values. Many of the plants are considered holy & sacred in our country like tulsi, peepal, mango, lotus, bael, etc. The leaves, fruits or flowers of these plants are used in worship or the plant itself is worshipped. The tribal people are very closely linked with the wild life in the forests. Their social life, songs, dances and customs are closely woven around the wildlife. Many animals like cow, snake, bull, peacock, owl, etc. also have significant place in our psycho-spiritual arena and thus hold special social importance. Thus biodiversity has distinct social value attached with different societies.

The biodiversity has ethical values based on the concept of 'Live and Let Live' which involves ethical issues like all life must be preserved. If we want our human race to survive, then we must protect all biodiversity, because biodiversity is valuable. People from far and wide can enjoy the aesthetic value of biodiversity and this type of tourism is known as eco-tourism. The willingness to pay concept on such eco-tourism gives us even a monitory estimate for aesthetic value of biodiversity. Eco tourism is estimated to generate about 12 billion dollars of revenue annually. India is one of the 12 mega diversity countries in the world.

Human being is the most dominant animal in the biosphere. He depends on the natural world of plants, animals, soil, air and water that existed since ages. Also, he created a world of social institutions, and infra structure using science, technology and political organizations. Both these are essential for the survival of the human beings. A combination of certain physical features and mental abilities have made human more powerful than any other animal. Early humans used natural resources to satisfy their needs with air, water, food and shelter termed as natural needs. With the dawn of industrial revolution humans have acquired the capacity to change their life styles towards more comfort. These acquired needs and greed coupled with shortsightedness have started to disturb the delicate ecological balance by depleting and degrading the vital life like supporting resources such as air, water and land.

The air we breathe, the water we drink, and the food we eat has all been badly polluted. Our natural resources are just dwindling due to over exploitation. If growth continues in the same way, very soon, we will be facing a 'doomy day'-as suggested by Meadows *et al.* (1972) in their world famous academic report 'The limits to growth'. This is unsustainable development which will lead to a collapse of the inter- related systems of this earth.

Environment belongs to all, and is thus important for all. Environmental studies is very significant since it deals with the most mundane issues like safe and clean drinking water, hygienic living conditions, clean and fresh air, fertile land, healthy food and development that is sustainable.

The United Nation's conference on Environment and Development held in Bio de Janeiro in 1992, and popularly known as 'earth summit' followed by World Summit on sustainable development at Johannesburg in 2002, just 10 years after the first summit have highlighted the key issues of global environmental concern and have attracted the attention of the general public towards the deteriorating environment. There is a Chinese Proverb "if you plan for one year, plant rice, if you plan for I0 years, plant trees, and if you plan for 100 years, educate people." If we want to manage our planet earth, we have to make all the persons environmentally educated. This is because; we are a part of the complex network of environment where every component is linked to another.

Using appropriate technology, reduce, reuse, recycle approach, prompting environmental education and awareness, resource utilization as per carrying capacity are some important measures for sustainable development. Environmental awareness needs to be carried through formal and informal education to all sections of the society. Every one needs to understand because environment belongs to all. By the following methods, we educate the people about environmental protection.

Among Students Through Education

Environmental education must be imparted to the students right from the childhood stage. It is a welcome step that now all over the country we are introducing environmental studies as a subject at all stages including school and college level, following the directions of the Supreme Court.

Among the Masses Through Mass-Media

Media can play an important role to educate the mass on environmental issues through articles, environmental rallies, plantation campaigns, street plays, real eco-disaster stories and success stories of conservation efforts. T.V. Serials like virusat, Race to save the planet, heads and tails, Terra-view captain planet and the like have been effective in propagating the seeds of environmental awareness amongst the viewers of all age groups. Among the planners, decision makers and leaders:

Since this elite section of the society plays the most important role in shaping the future of the society, it is very important to give them the necessary orientation and training through specially organized workshops and training programmes. Voluntary organizations can also help by advising the government about some local environmental issues and at the same time interacting at the grass root levels. They can act as an effective and viable link between the two.

Value based Environmental Education

The fundamentals of environment can really help to create a sense of earth citizenship and sense of duty to care for the earth and its resources and to manage them in a sustainable way so that our children and grand children too inherit a safe and clean planet to live on. The environmental education can be made value based by the following:

1. Preparation of text books and resource materials about environmental education can play an important role in building positive attitude about environment.
2. Social values like love, compassion, tolerance and justice which are the basic teachings of most of our religions need to be woven into environmental education.
3. Cultural and religious values enshrined in Vedas like *"Dehi me dadamite"* i.e., 'you give me and I give you' emphasize that man should not exploit nature without nurturing her. Our cultural customs and rituals in many ways teach us to perform such functions as would protect and nurture nature and respect every aspect of nature, treating them as sacred, its rivers, earth, mountains or forests.
4. Environmental education should encompass the ethical values of earth centric rather than human centric world view.
5. Global values stress upon the concept that the human civilization is a part of planet as a whole and similarly nature and various natural phenomena over the earth are interconnected and inter-linked with special bonds of harmony.
6. Spiritual values highlight the principles of self-restraint, self-discipline, contentment, reduction of wants, freedom from greed and austerity.

The above mentioned human-values, socio-cultural, ethical, spiritual and global values incorporated into environmental education can go a long way in attaining the goals of sustainable development and environmental conservation. Value based environmental education can bring in a total transformation of our mind-set, our altitudes and our life styles. The value elements in environmental education alone can succeed in achieving- the real goals of environmental literacy. The role of every individual in preventing pollution is of paramount importance because if every individual contributes substantially, the effect will be visible not only at the community, city, state or national level, but also at the global level as environment has no boundaries. It is the responsibility of the human race which has occupied the commanding position on this earth to protect the earth and provide conducive environment for itself and innumerable other species which evolved on this earth. A small effort made by each individual at his/her own place will have pronounced effect on the global level. It is aptly said *'Think globally act locally.'*

PRESERVE ECOLOGY: TEACHINGS OF JESUS AND ST. FRANCIS OF ASSISI

Lilly V.S.

"In the beginning, God created the heavens and the earth" (Gen.1/1). After creating the heavens, the sea, the earth and all it contains, God created man and woman. All forms of life, the living and the non-living speak of God's relationship with the earth. "The spirit of God was moving over the face of the water" (Gen.1/2). God entrusted the whole of creation to the man and woman (Gen 2:3). Made in the image and likeness of God, Adam and Eve were to have exercised their dominion over the earth (Gen 1:28) with wisdom and love. Instead, they destroyed the existing harmony by deliberately going against the Creator's plan, that is, by choosing to sin. This resulted not only in man's alienation from himself, in death and fratricide, but also in the Earth's "rebellion" against him (cf. Gen 3:17-19; 4:12).

But God who is merciful and kind hears the cry and redeems the earth. Jesus Christ comes to redeem the whole universe from its curse and he redeems it at his crucifixion. Creation was thus made new (cf. Rev. 21:5). The birth and life of Jesus is framed within the widest context of cosmic history. He is active in bringing forth creation; through him the universe, the earth and all life was created (John 1:3-5). Hence all of these crucial moments in the emergence of the universe have a Christic dimension.

The Redeemed Earth Cries Louder than Before

Today, the world is in the throes of both ecological and economic crisis, which are already assuming catastrophic dimensions for human kind and for the rest of the environment. The manifestations of these crises include air, water, and soil pollution through toxic chemical and radioactive waste, soil erosion, salinization, desertification, deforestation, the extinction of plant and animal species, global warming and loss of human life and spirituality. Many of the victims of this human spiritual destruction are destined to become alcoholics, drug addicts and criminals.

Pope John Paul II tells that the human creature receives a mission of government over creation to make all its potential shine. His responsibility, defined in the book of Wisdom, is to govern "the world in holiness and justice" (wisdom 9:3). But humanity has disappointed the divine expectation. Above all, in our time, man has unhesitatingly devastated wooded plains and valleys, polluted the waters, deformed the earth's habitat, made the air unbreathable, upset the hydro geological and atmospheric systems, blighted green spaces, implemented uncontrolled forms of industrialization, humiliating the earth, that flower-bed that is our dwelling. It is necessary, therefore, to stimulate and sustain the ecological conversion.

Faced with the widespread destruction of the environment, people everywhere are coming to understand that we cannot continue to use the goods of the earth as we have in the past. Today, the rivers and oceans, which give life, are used as dumps for our wastes, the forests are destroyed for crass commercial gain, and the soil is squandered for profit, to benefit the few. The environment became polluted and contaminated when man started exploiting natural resources for his development. This has resulted in the present dangerous situation in which the existence of even human beings is difficult. We are alienated from ourselves, from one another and from nature. Man violated the Divine regulations and transgressed the measure set by God.

Certain elements of today's ecological crisis reveal its moral character. First among these is the indiscriminate application of advances in science and technology. Many recent discoveries have brought undeniable benefits to humanity. Unfortunately it is now clear that the application of these discoveries in the fields of industry and agriculture have produced harmful long-term effects. This has led to the painful realization that we cannot interfere in one area of the Ecosystem without paying due attention both to the consequences of such interference in other areas and to the well being of future generations.

In Search of a Solution

Theology, Philosophy, and Science all speak of a harmonious universe, of a "Cosmos" endowed with its own integrity, its own internal, dynamic balance. This order must be respected. The human race is called to explore this order, to examine it with due care and to make use of it while safe guarding its integrity. So, an education in Ecological responsibility is urgent. A true education in responsibility entails a genuine conversion in ways of thought and behaviour. Our attitude must be constructive and harmonious. We have to understand the earth as a gift or blessing from God, entrusted to our care, in order to experience and feel Gods' loving presence, kindness, generosity, power, beauty and glory.

God-Man-Nature Relationship in Different Religions

India's religious and cultural heritage will inspire modern man to change his attitude to nature if he takes courage to dive deep down into the recess of God-Man-Nature relationship as portrayed in the Sacred writings of Christianity, Hinduism and Islam.

In the Bible

The Old Testament begins with the story of Creation. God said, *"Let there be light and there was light"* (Gen.1:3). This indicates that the universe has come into existence from the mind of God. The main theme of *The Bible* is the restoration of humanity and of the whole creation to its original harmony. God's compassion for the creatures is revealed in the following verses of Psalm 145: *"The eyes of all look to thee, and thou givest them their food in due season. Thou openest thy hand, thou satisfiest the desire of every living thing"* (15-16). During the journey to the promised land, the Lord walks before the Israelites by day in a pillar of cloud to lead them along the way and by night in a pillar of fire to give them light (Exod: 13:21). When man's spirit is redeemed, creation itself will be set free from bondage. Therefore, all are in need of inner transformation. They will then activate all their resources and respond to the crisis and heal the earth by eliminating inequities, injustice, oppression and greed, which are the primary causes of the environmental crisis. If the Bible begins with the creation of "heaven and earth" (Gen.1:1) it ends with the new creation of a "new heaven and a new earth" (Rev.21:1).

In the Gita

The Vedic men considered Ecology as a Sacred Science. Hindus believe that God is the Creator and only God has absolute sovereignty over all creatures including man. Lord Krishna says to Arjuna:

> Of the whole (wide) universe
> The origin and the dissolution too am I (The Bhagavad-Gita 7:6).

So, man has no dominion over his own life or non-human life. According to *The Gita* the seeker of Truth is blessed with a new vision of reality when he overcomes his ego-centeredness and becomes liberated from greed. *The Gita* advocates the three-fold 'marga' (way) of spirituality, viz, 'the jnana marga', 'bhakthi marga' and 'karma marga'. *Jnana* enables one to obtain a holistic vision of reality while *bhakthi* leads one to a reverential realization that one is a part of the totality. *Karma* helps one to have creative participation in the divine work of nourishing the entire universe and restoring its lost ecological balance and harmony.

In the Quran

According to *The Quran,* Allah is the ultimate cause of the whole creation. He is the beginning and the end. Before the creation of man, the earth was clean, pure, beautiful, healthy and rich. She lost her natural virginity when the human species emerged on earth. The environment provided by nature was perfect and it became polluted only when man tried to change it. Man disturbed the balance, which was not permitted by the word of God as revealed to the Prophet Mohammed. *The Quran* provides solution to the pollution problems. Man has to repent and purify himself. He has to keep himself away from the ignoble and destructive actions, which lead to pollution in society because "God does not wish injustice to the creatures of the world. For to God belongs all that is in the heavens and the earth, and to God do all things return" (The Quran 3: 108-109).

Ever since the beginning of the world, Nature has been considered as one of the popular means to experience God. Nature is nearer to God and man makes use of nature to reach out to God, the absolute and the infinite. This we have already seen in the heritage of the three major world religions viz, Christianity, Hinduism and Islam. An Ecologist experiences the invisible God through the visible nature. For him every blade of grass, every sweet smelling flower, every star in the sky, every bird of the air, the running brook and the silent mountains are the visible manifestations of God, the Creator, the Protector, the Redeemer, the Artist and the Poet.

Jesus, the Ecologist

Jesus is a lover of nature. Nature becomes the background of His life and death. The gospels tell us that nature played an important role in Jesus' life. At his birth, Luke tells us that 'he was laid in a manger, because there was no place for them in the inn' (Luke 2:7).

Jesus grew up in a rural environment and had an intimacy and familiarity with a variety of God's creatures and the process of nature. Nature, which consists of flocks of sheep and shepherds; the manger and the pastures, the sky and the stars. The Sea of Galilee, the hills and dales of Jerusalem, all seem to be dear to him and so he frequently visits them. He goes to the mountain to address the crowds. He knows the plants, trees, flowers, and fruits, the birds and animals of the locality. Being a son of the soil he is in close contact with all that the Palestinian landscape provides.

It was during his sojourn in the desert that Jesus came to accept and appreciate the messianic ministry he was about to embrace. In order to be fully open and receptive to his call, Jesus forsook the company of people and spent time with the wild animals in the wilderness. He regularly returned to the hills to pray and commune with the Father (Mt 17:1; Mark 6:46; Mt 14:23), especially before making important decisions like choosing the disciples (Luke 6:12). His teaching ministry was not carried out in buildings, in synagogues or in the temple, but in the cathedral of nature.

Dr. R.J. Raja says that all nature in its varied splendour take part in the teachings of Christ. The sun and the rain (Mt.5: 45), the scorching heat and the south wind (Lk.12: 55) the clouds and the showers (Lk.12: 54), the earth and the sky (Lk.12: 56), the flash of lightning (Mt. 24:27), the rock and the sand (Mt.7: 26), the seeds and the grains (Mk.4:2-8), the lilies and the grass (Mt.6:28-30), the thorns and the thistles (Mt.7:16), the figs and the grapes (Mt.7:16), the moth and the rust (Mt.6:19-20), the sparrows (Mt.10:29) and the eagles (Mt.24:28), the ox (Lk.13:15) and the dogs (Lk.16:21), the fish and the serpent as well as the scorpion (Lk.11:11), the sheep and the goats (Mt.25:32). In the last supper the bread and vine becomes the flesh and blood of Jesus Christ. Mount Calvary becomes the Holy Mount with the crucifixion of Jesus. Finally in and through his death, Jesus participated in the most radical way in one of the key processes of nature.

St. Francis of Assisi as Ecologist

On November 29, 1979, Pope John Paul II declared Francis as "Patron Saint of Ecologists". After his conversion, Francis became one of the most famous preachers of peace and non-violence. Beyond preaching he expressed his conversion by his lifestyle; he moved from the comforts of the city of Assisi to the leprosarium on the periphery, he abandoned the selfishness of his youth for a life of itinerant preaching, and in his later years he spurned the popularity of his fame had brought him for the life of a mystic and hermit in the wilderness.

St. Francis perceived the sacramentality of Nature. He arrived at sympathy and synergy with all things. He was a marvelous poet, a poet of essence, one capable of grasping the sacramental message echoing from all things. Conversion did not kill the poet in Francis but energized him. St. Francis was able to personalize all his relations, because he saw them as sacraments of the divine presence; the lark is Sister Lark, the sun is Brother Sun and Master Sun, the earth is Sister and Mother Earth, robbers are Brother Robbers to one who runs after them to give them the rest of what they had not succeeded in taking.

An old legend that became a song that is popular in Umbria to this day gives a good idea of this inclusively of St. Francis's ecological love. One day Francis said to the Lord, weeping:

I love the sun and the stars,
I love Clare and her sisters
I love the human hearts
And all beautiful things, .
Lord, forgive me
For I should love only you.
The Lord smiled and replied:
I love the sun and the stars,
I love Clare and her sisters,
I love the human hearts

And all beautiful things,
My dear Francis
You need not weep
For I also love all this.

Francis shows no fear of our cosmic roots. He does not define human beings in terms of what distinguishes us from others but by what we have in common and what makes us kin with others.

The greatest Nature Mystic of the middle Ages was St. Francis of Assisi. He is the supreme example of how the mystical life expresses itself completely through a spontaneous and effective love for God's creation. To him, every visible and invisible creature was a theophany or appearance of God. He preached to his little sisters, the birds. He was happy to receive the service of the falcon. He enjoyed the friendship of the pheasant and soothed the captured turtledoves. Nature then became his intimate friend and everything in Nature responded to his fraternal love and sympathy. Thus, he could discover the sacramentality of Nature and acknowledge its fraternity. He saw God's sanctifying presence in every created thing-from Brother Sun to Sister Bodily Death.

In St. Francis's eye the earth and water, clouds and sky were symbolic. If the warming Sun was a smile of God, Brother Fire was God's light. The singing larks were a Holy Choir, God's little Orchestra of Joy. When wind was God's breath that cooled the world, blowing grasses were the unfurled banners of God. Mother earth was a symbol of God's bounty. He could feel God's presence in the leaping young doe and see God's tears of joyous weeping in the rain. Once he preached a sermon for the birds, his feathered friends.

Francis valued the diversity and beauty he saw in nature. Celano wrote that Francis insisted his brothers leave a border around the community garden untouched so that wild grasses and flowers could announce the beauty of the Creator. He forbade his followers to cut down a whole tree so that it might have hope of sprouting again. Francis loved the creation because in it he found the Lord, but his love was not limited to popular or beautiful species. When he found an abundance of flowers, he preached to them and invited them to praise the Lord as though they were endowed with reason.

During his last few years he spent over half his time in the wilderness deepening his experience of contemplative prayer. He withdrew with a few companions to seek God in nature, and it was there that he had his most profound mystical experiences, most notably the stigmata. The hermitages in which he dwelled were really no more than small caves or huts constructed of sticks, leaves, and mud, but they were located in areas of great natural beauty.

During this twilight of his life on Earth, Francis wrote his greatest piece of poetry, "The Canticle of the Creatures". This poem reveals the summit of the spiritual journey of Francis as a lover of nature. As Francis sang of the diverse parts of nature, he described them in remarkably intimate terms; he expresses great joy at having lived in relationship to them. He wrote:

> Praised be You, my Lord, by Brother Fire,
> Through whom you light the night,
> And he is beautiful, and playful and robust and strong.
> Praised be You, my Lord, for our Sister Mother Earth,
> Who sustains and governs us,
> and who produces varied fruits with
> coloured flowers and herbs.

In this poem, Francis invites all creation to praise its creator. He praises God through the images derived from nature. This magnificent Hymn expresses his mystical vision. He becomes so intimate and familiar with the wonders of creation that instead of losing himself in the vastness of the created world, he embraces them as "Brother", "Sister", "Mother", etc. In the view of St. Francis the human race cannot live without the Lord's creatures and it is with these creatures that we greatly offend its Creator.

Francis–a Preacher of Reconciliation with the Nature

The Little Flowers of St. Francis relates the story when Francis visited the town of Gubbio while a huge, ravenous wolf was terrorizing it. Francis, filled with faith in God, journeyed out of the town, found the wolf, and began to preach peace to it. The saint commanded it to stop terrorizing the citizens, pledging that they would feed him and care for his needs. He then led the wolf into the town, preached penance and peace to the citizens, and forged a covenant between the humans and the wolf, bringing about happy reconciliation and peaceful lives.

He was understood to be a man of peace and a preacher of reconciliation. He wanted the residents of Gubbio to be free from their fear of the wolf, so he exhorted them to reform their lives and live out gospel values. Francis did not seek to destroy or eradicate the wolf; he sought to become its brother and bring it into peaceful relationship with its environment. For our peace and well-being, we humans must undergo conversion and strive for reconciliation with God, each other, and Nature.

Conclusion

India is famous for her rich and diverse religious and cultural traditions. She has a rich resource of beliefs about Nature and rules for the sustainable utilization of material resources. But today Indians have up-rooted themselves from the earth and embraced rootless consumerism. Success and physical attractiveness are considered as the twin gods of consumerist culture. What we mean by 'success' today is totally different from the perception of Christ, Lord Krishna and Allah. The successful person is one who loves his brethren as Christ loved us.

The environmental ethics that emerges from the present environmental crisis encourages us to preserve, protect, and conserve Nature. If we are absorbed in the concerns of consumerist culture, which is rootless, and spiritually bankrupt we ourselves become rootless and float on the surface of life's ocean, without realizing our own potential for spiritual growth and relationship with others. Hence we must have a new vision of God, Man and Nature. From this new vision a new Global Theology or Eco-Spirituality may emerge.

Eco-Spirituality demands the art of being in communion with nature, which is in fact opposed to domination and mastery over nature. It is opposed to Ego-Spirituality and so it is in resonance with and responsive to the creative cosmic energy that is flowing constantly through nature. Human beings must be open to this flow of energy that energizes them and renews them. Eco-Spirituality opposes violence. So, *non-violence* is its hall-mark. When humans love Nature as they love themselves, non-violence and gentleness will become the only way to survival. Therefore like the Ecologists-Jesus Christ and St. Francis of Assisi, humans must strive to make Mother Earth rise up with all her awe - inspiring resources to the primal dignity she enjoyed when God said, "it was good" (Gen.1: 4-10).

Bibliography

Albert J. Lachance and John E.Carroll, *Embracing Earth.*

Ali Ahamed (trans.), Al-Quran. Delhi, Oxford University Press, 1987.

Apostolic letter, Inter Sanctos, A AS 71, 1979.

Cox, Michael, *A Handbook of Christian Mysticism*: 1983.

Rev. Ed. Great Britian, Crucible, 1986.

Sr. Sheila Kannath, CMC: *The Groaning Creation* (Eco-spirituality), 2001 Mercy Publications, Palakkad.

Raja R.J. (1997), Eco-Spirituality, Bangalore, N.B.C.L.C.

Leonardo Boff, Cry of the Earth, Translated by Philip Cry of the Poor Berryman.

Pope John Paul II (1990), Message given for the celebration of the World Day of Peace: January 1, 1990.

Pope John Paul II, General Audience Address, January 17, 2001, (Ecological Conversion).

Sean McDonagh, SSC, Ecology and Religion.

The Bhagavad Gita (trans.), 1962; rpt. England Penguin Classics, 1978

The Holy Bible, Revised Standard Version (Catholic Edition for India) Bangalore; Collins 1980.

RELIGIOUS FAITH TO PRESERVE ECOLOGY

S.Thadedus Maria Ignatiammal

The alarming ecological imbalance, which has shocked humanity with serious disasters has led man to introspect as to what has gone wrong in the cosmic set up. The living organisms and their environment, the two basic components of nature are interdependent and mutually reactive. Man and nature had a harmonious co-existence. Slowly man began to detach himself from nature. Progress in agriculture and industry became the sole criterion for the development of any nation. Though the history of development is based on the "green pastures" of rich and developing nations, it is an irony that the very progress has degraded nature and polluted the environment. Man's greed for power and glory has resulted in unlimited exploitation of every bit of natural resource. Humanity has conveniently forgotten that this splendid nature is a rich heritage, which has to be conserved carefully for the benefit of the future generation. This sophisticated exploitation of nature has not only threatened the very survival of man but also the natural habitat of flora and fauna. This threat to humanity is being viewed seriously and the lovers of ecology are trying to create sincerely an awareness to bring back the pristine glory of nature. Religions have started responding to the environmental challenge as the world has become hotter, stormier, more violent, unsettled and less bio-diverse. They have contributed their mite to the struggle to secure the earth community's well being. They have also reexamined elements of scripture and tradition to refocus the affirmations and ethics in ecological - alert terms.

God and Nature in Religious Traditions

Forests are central to the civilization, culture and faith in India. Most of the South Asian religions are rooted in the forests. The Christian tradition can be traced from the Book of Genesis. The story of creation delineated there very clearly shows that man and nature had a very harmonious co-existence. God was very meticulous in planning the whole creation and placed man in an un-polluted surrounding. The Garden of Eden gave the required ambience for man to live in peace. Genesis 1:20 says that man has 'dominion over' the living things. That means that man should be a conscientious steward to carry out the God given responsibility of caring and nurturing all his creations. God saw that the earth, teeming with life in all diversity was good. The living earth is given to the humans to rule over it. But then rule does not mean the license to destroy but the responsibility to protect, foster and nurture. The 'dominion over' can be taken as the permission to use the garden for food. Ultimately what this phrase means is that man should have friendly approach to nature "where lamb, wolf, children and serpents, cow and lion peacefully live together". This explains the peaceful co-existence of man and nature in the Christian tradition.

In the New Testament too, there are allusions to man and nature. God who takes care of the lilies also protects man. Many of his parables make reference to the plants, birds and animals. This shows Jesus' concern for the diversity in nature. So, on the whole, indiscriminate use of nature and irresponsible motives of the Christians to go against the wish of God are rampant, because the Christians have grossly misunderstood the words "to fill the earth and subdue it" (Gen 1:28).

The Hindu faith also expresses the same sentiment on the lines of Christian thinking. It says that the human beings have forgotten that a particular natural condition on earth enabled life to come into existence and evolved to the human level. Humanity is disturbing this natural condition on which his existence along with the existence of all other forms of life depends. According to the Hindu scriptures dharma sustains all species of life and helps maintain harmonious relationship among them. If any one disturbs ecology, it is adharma. "Conserve ecology or perish" is the message of the Bhagavad-Gita. To sum up, Hindu philosophy asserts that life without contribution towards the preservation of ecology is a life of sin and life without specific purpose or use.

In Islamic tradition too, The Quran says that Allah who created all things, have some regulations to respect all creations, So going by rules all should learn to respect nature (Q. 7.55-57) Any one who destroys nature, disturbs the creations of Allah, he will also be punished and put into hell. (Q. 4.117-119).

• Zorastrianism too harps on the idea that only through nature man can attain redemption. It invites every one to have a life inter-connected with nature. It also insists that the duty of the Zoroastrian is to preserve the seven elements namely air, water, earth, plant, animals, man and fire. Their worship is centered on these seven elements of nature.

Man's Callous Attitude to Nature

The various religious motives presented above picture the respect man has towards nature. But man, because of his greed has indulged in the indiscriminate use of nature. In this context, Lynn White in his article "The Historical Roots of our Ecological Crisis" says that the western world's attitude towards nature was shaped by the Judeo-Christian tradition. This tradition involved the concept of a world created solely for man. White writes, "God planned all (of creations) explicitly for man's benefit and rule no item in the physical creation had any purpose save to serve man's purposes. We have already seen in the older religious traditions humans are presented as part of nature rather than the rulers of nature. Animistic religions established that there is a spirit in every tree, mountain or spring. But some how or other the Western Christianity in contrast to the Eastern religions and paganism separated humans from nature and encouraged exploitation of it for our own ends and added that it is God's will.

The Bible says, "God created the heavens and the Earth", (Genesis 1:1) But man who has taken the control from the Lord of the earth has started using the earth to his own selfish ends. Today through large scale industrialization the atmosphere has been badly polluted and most of the land has been turned barren through indiscriminate commercial agriculture. Deforestation is a major environmental problem today. Tropical forests are being cleared for the construction of dams, approach roads, residential quarters and for storage of building materials. Irreversible fragmentation of forest and the gradual degradation of the remaining vegetation also affect the flow of rivers. Man's indiscriminate felling of trees radically affects the very life and culture of the tribals. The tribal population gets displaced. The tribals are not only uprooted from their homeland but from their very culture and faith. They have Gods and Goddesses in their homeland surrounding the forests that cannot be re-shifted or resettled. Artificial monoculture plantations lead to the loss of biological strength, Lack of availability of herbal and traditional medicines further deteriorates the health of the forest dwellers. Biospheric potential and underground water pockets have disappeared.

God made earth and put forth vegetation (Gen 1:11). But the modern development policies and projects result in desertification of land which in turn becomes arid and lost its nutrition. The earth is completely depleted. God spirit was brooding over the face of the waters (Gen 1:1). But water is the most polluted element on earth today. In India ground water resources are equivalent to 10 times its annual rainfall. Due to the adding of tube wells, the scenario has changed, leaving the dug wells of the poor dry. Nearly 70% of the available water is polluted. Slow and steady death of Dal Lake due to pollution threatens the livelihood of 50,000 fisher folk living there. Ganga with its tributaries is the largest and very important river basin, has been a symbol of purity is in utter disregard to its sanctity. About 132 industrial plants dump their waste into the river. New Delhi's sewage system dumps more than 200 million litres of untreated waste into the Yamuna River per day. Kerala, a southern state has a notoriously poor record for checking the industrial pollution of its waters. Mass fish kills are reported from all the rivers that flow adjacent to industrial areas in Kerala. The major rivers of Karnataka, The Cauvery, and Tungabhadra are contaminated with high density of CO_2 emitted by the nearby industrial establishments. Modern fishing techniques, and over fishing has resulted in water pollution and extinction of marine wealth, threatening the very survival of the traditional fisher folk.

Air pollution resulting from gaseous emissions from industry, thermal power stations, automobiles, domestic combustion etc completely corrodes the environment. To some extent increase in CO_2 concentration in atmosphere increases the photosynthetic rate and consequently results in plant growth, acting as fertilizer especially in hot tropical climates. When it is in higher concentration the thick layer of this gas functions like the glass panels of a green house allowing the sunlight to filter through but preventing the heat from being re-radiated into outer space. This is the so called green house effect. The net result is the heating up of the earth's atmosphere CO_2 increases the temperature by 50% while CFC's are responsible for 20% increase and the polar ice caps melt. If all the ice on the earth started melting 200 feet of water would be added to the surface of all nations and low lying coastal cities. Green house effect leading to occurrence of hurricanes and cyclones, snow melts in mountain causing more floods during monsoon. If unchecked it would alter temperatures, rainfall and sea levels of earth.

Oil and coal combustion releases sulphur and nitrogen oxides into the atmosphere mixed with rainwater, it then falls as acid rain. Acid rain increases soil acidity, thus affecting flora and fauna. It also corrodes buildings, monuments, statues, railings etc. As a result of acidity heavy metals in water increase beyond safe limit. The lakes become unproductive, and have become graveyards to the fish.

The Ozone layer in the atmosphere is nature's umbrella and protects us from the harmful U.V. radiation from the sun. The chlorofluro Carbons, methyl chlorides, nitrous oxides and methane react with Ozone layer and deplete it. Every 1% decrease in the concentration of Ozone will lead to 5% increase in the number of non-malignant skin cancers and 1% increase in malignant cases. U.V. radiation will lead 25% decrease in the yield of many crop fields. From the above descriptions we can infer that man is not eco-friendly and his callous attitude reveals that he has completely forgotten the divine presence in nature.

What is the Need of the Hour?

When the whole world is facing a crisis because of the ecological imbalance as Christians what is our responsibility? Are we supposed to remain passive and become victims of the pollutants created by man's selfish greed? The answer is a firm 'No' and we should come out with clear-cut solutions. At the outset the theologians should come away from man-centered theology and direct their deliberations towards eco-centered theology. We should mobilize people to protect the environment because spirituality is closely linked to nature. The Director of Zimbabwean Institute of Religious Research and Ecological Conservation, Mr. Soloman says that holding services out doors and baptizing their members in running rivers will strengthen their dependence on and respect for the environment.

Peter Harris, Director general of Rocha International, a Christian conservation group, wailing over the damage done to the environment says, "Environmental stewardship is at heart of Christian faith, and it is a tragedy that many Christian groups, particularly western world have completely forgotten this". Only by protecting natural environment the universe can experience eco-justice prosperity and unity. People should understand that nature reflects the qualities and attributes of God and should therefore, be greatly respected and cherished.

We need to build a "society of decrease" based on the 6R programme given by United Nations Conference on Environment and Development at Rio in 1992. They are Re-evaluate, Restructure, Redistribute, Reduce, Reuse and Recycle. In the context of eco-system, we must try to bring out a new image of human life, of well being, of growth educating ourselves to limit our use of material things in order to have access to nonmaterial goods. Once our greed for material things is under control and primary needs satisfied the unlimited growth of the non-material goods of culture, civilization and spirituality can be at the center of our human experience. Another important consideration is that we should remember that the minimal use of the material goods will save the world for the future generations. We should also have the basic feeling that the future generation has a title to the earth and the cosmos, to resources of nature just as we do.

The Christian theologians should create a Biblically-informed imperative to pursue a life, which is ecologically fitting and socially just. In this context I would like to quote what Pope John Paul II has said in one of his letters. He asserted that unpolluted environment is the basic right of human beings. This valuable concept should be followed in letter and spirit. An ecological reformation is the need of the hour. We need a reformation because as Christians we have failed in sustaining the eco-balance. As Dieter T. Hessel rightly observes, "An ecologically reformed Christian theology will reinterpret basic doctrinal themes in ways that integrate ecological insight and value and reconceived Christian ethics to encompass human relationships with other beings in the bio sphere." An ecology oriented worship and liturgy will make people take an undivided interest in the sustenance of the biosphere. To conclude, I quote Jane Blewette who says that eco-justice and social justice are the two sides of the same coin. If we say that we work for the development of the people and remain passive to the massive destruction of nature, we are deceiving ourselves.

Bibliography

Anthony and Miriam Hanson, *A Commentary on the Book of Genesis.*

Nalunnakkal, George Mathew, *Green Liberation Towards an Integral Ecotheology.*

Wiolenga, Bas, *Towards an Eco–Just Society.*

Lean, Mary, *Pollution and the Environment.*

Sharma, P.D., *Ecology and Environment.*

Smithsomian Institution Symposium, *The Quality of Man's Environment.*

Hessel, Dieter T., *Christianity and Ecology.*

SCIENTIFIC AND RELIGIOUS TESTING OF FAITH: WRESTLING WITH MYSTERY

Pragyesh Agrawal & Varsha Gupta

> "Faith is to believe what you do not see; the reward of this faith is to see what you believe." – *St. Augustine*

Before entering into the intricate job of testing faith let us first try to define it. Faith can be defined[1] as follows:

- The idea that something can be made true, merely by wishing it to be so.
- The proposition that something is true, even if there is no evidence to support it.
- The belief that something is true, in spite of evidence to the contrary.

According to Lord Buddha, "On a long journey of human life, faith is the best of companions; it is the best refreshment on the journey; and it is the greatest property." It can also be defined[2] as confident belief in the truth, value, or trustworthiness of a person, idea, or thing. We all go through our life believing that if our faith is strong enough, nothing will harm us. God will protect us overall; all will be fine and perfect. How can people so blindly believe in something without proof? Faith does not save us. Faith is not a cream we apply to protect ourselves from something. Faith is belief and trust. In the words of Mahatma Gandhi, "Faith is not something to grasp, it is a state to grow into."

The key to understanding faith lies not in faith itself but in the object of faith: God. Faith is only as good as who you put it in. Faith in a false god is useless and it does not save you. But faith in the true God *does* save you. Faith is only as good as who you put it in. Faith in the true God results in salvation. Faith in a false god results in damnation. Finally, faith "works" because of who it is placed in. It is not a thing dependent upon you for its vitality. It is not made real by us. It is made real by *who we place it in* that is why true faith in the true God is independent of your circumstances.

It is said that God gives a measure of faith to everyone. During trials and tribulations do you withdraw it from your spiritual walk, whine and complain, only to be faithful again when things get more comfortable? Or, do you continue on in faith (because it is in the Creator of the Universe), in spite of your circumstance, and do exactly what scriptures say? To live by faith means that you continue trusting God because our faith is in HIM and not in your comforts, your health, your needs, your emotions, or your intellect. To live by faith means that we trust the Lord in all circumstances. In this way, God is glorified. Religion may be a survival mechanism. So are we born to believe? So, in all religion is a matter of faith and hence all Gods are a matter of faith. When everything is a matter of faith, can there be right and wrong?

Testing Faith

"Faith is knowledge within the heart, beyond the reach of proof." –Kahlil Gibran

Did you ever try to ***really*** test your faith? Or have you ever wondered where God is in the midst of our difficulties even though we have been crying out to Him for deliverance and have seen none? Have you ever found yourself with nowhere to turn, no real answers, no light at the end of the tunnel? Have your hearts and minds been filled with anguish, doubts, fears, and commotion and faith seems more like a distant concept than present reality?

In our everyday experience, just about everything seems to have a beginning. In fact, the laws of science show those even things, which look the same through our lifetime, like the sun and other stars, are running down. The sun is using up its fuel at millions of tons each second. Since, therefore, it cannot last forever; it must have had a beginning. The same can be shown to be true for the entire universe. So do we need to have a beginning of testing our faith in order to keep on growing in it?

Faith and the utter belief in something lacking in proof scares people at times. It becomes such a self-involved thing, 'God will love me if I just believe in him', and it is a cycle that goes on till the believer simply lives in a bubble of blind faith. What happens when someone pops that bubble? When they enter in to the real world and realize that it is not as wonderful as they thought", how can their faith help them then? Many people, including many important and well-respected scientists, just do not want there to be anything beyond nature. They do not want a supernatural being to affect nature, no matter how brief or constructive the interaction may have been. In other words they bring on a priori philosophical commitment to their science that restricts what kinds of explanations they will accept about the physical world. Sometimes this leads to a rather odd behavior.

The crux of the matter is this if one accepts there is a God who created us, then that God also owns us. He thus, has a right to set the rules by which we must live. In the Bible, He has revealed to us that we are in rebellion against our Creator. Because of this rebellion called sin, our physical bodies are sentenced to death - but we will live on, either with God, or without Him in a place of judgment. One might say, 'But that means I have to accept this by faith, as I can't understand it.' It is important to remember that for a theory to become a law, it must work in all possible cases, and that only one case is required to disprove a theory. The scientific method yields closer and closer approximations to the truth. Science is not a body of knowledge. It is a process and a way of thinking. History has shown that what is accepted, as scientific fact during one era is likely to be proven false by scientists of the next era.

Almost every predominant theory of the world is not discarded entirely after its age has passed, but contained within the next theory as a special case. Before Newton, it was thought that objects have a fundamental desire to be at rest, and that if set into motion they would, for that reason, gradually return to that state. Newton considered that picture of the world, that explanation of the why, to be false, clearly false, in all circumstances. But the *prediction* that that picture gave - and this is the fundamental part of the theory, the only part that matters to Science - remains *true* for everyday circumstances. Roll a ball along a table and unless you have some frictionless surface available, it will slow and then stop at every single time. Similarly, the planetary motions explained by Kepler remain largely intact. The picture of the world that Einstein's theory of relativity gave may have overturned the Newtonian world-picture everywhere in every circumstance, but its description of events only overturns Newton's in relatively exotic situations[3] and that is all that science is really concerned with. What is worth observing here is that faith and facts may not go together in all cases Faith sometimes can dominate facts.

Blind Faith or Tested Faith

It has been stated over and over again that blind faith in religion is not only unhealthy for an individual, but also for other people around that individual and indeed for that religion itself. It can be concluded that blind faith in scientific theory is just as harmful. There seem to be a lot of undereducated individuals who believe that science is infallible. History has shown that what is accepted, as scientific fact during one era is likely to be proven false by scientists of the next era. "Science"-at the time, synonymous with religion, of course - once stated that the world was flat. The Earth was once the center of the universe.

Many people believe unquestioningly in Einstein's laws of motion. Einstein himself once stated that the largest number he could comprehend was one. Sure, his laws work more or less universally for everything we have figured out so far, but they may be disproved tomorrow. The voyager satellites are moving a lot slower than they should be, and no one seems to be able to figure out why. Further, we know for a fact that the current atomic model is false. We just have not figured out a better one yet. Numerous people want to test their faith in religion and science as a sole candidate comes forward to help them out.

It can be said that "Reality that science cares about is testable, material reality." Anything else, science doesn't presume to have answers on. A scientist will try to figure out all the "hows", and the "whys". When "why" means, "what caused this". But as for finding the "meaning" of things, that is not something science cares about. All faith will do is force them into either another bubble of complete denial or one of rebellion. Obviously there has to be more to life than this. How can we truly believe that we are the beginning and end of life? Believing in that is as bad as blind faith. This is what life is about, finding things out for you, trying to gain as much information on everything as we can. Don't believe blindly in God; think about your faith.

Think about this one: *What truth?* How do we know that there is a truth? Science is very good at suspending judgement on something, saying, "This is the best we have got". But the criteria for "better" must fall back on the method itself. And it would be circular for the scientific method to test itself. Skeptics have focused on the examination of paranormal claims. They do not deal with religious claims *per se*, unless they can be examined empirically. Secular humanists, on the other hand, do wish to deal with religious claims, testing them as best they can. Interestingly, in recent years the borderlines between the paranormal and religion have blurred and it is often difficult to tell when we are dealing with paranormal or religious phenomena. Thus spiritualism, near-death experiences, and communication with dead people interest both paranormal and religious investigators. Similarly for the appeal to intelligent design - a classical philosophical argument - now introduced within evolutionary biology and cosmology.

It would be more appropriate if we use the term *paranatural* to refer to religious claims that are capable of some empirical resolution and are not transcendental or supernatural. In this sense they are similar to testable paranormal claims. A good example of the overlap is in the popular belief that mysterious intelligent and beneficent extraterrestrial being are visiting earthlings and floating them aboard spaceships. This is a quasi-religious phenomenon reminiscent of angels, and other divine or semi-divine beings of earlier ages.

In order to analyze the relationship between science and religion, we need to define and characterize each domain. Many consider that religion offers a special kind of higher spiritual truth. They maintain that there are *two* truths:

- The truths of science, employing the methods of scientific inquiry, and testing claims empirically, rationally, and experimentally; and

- the truths of religion, which transcend the categories of empirical fact and logic.

Skeptics are rightly dubious of this latter claim

The most reliable methods, they insist, are those that satisfy the objective standards of verification and justification. The historic claims of revelation in the ancient sacred texts are insufficiently corroborated by reliable eyewitnesses or are based upon questionable oral traditions. These were compiled many decades, even centuries, after the alleged death of the prophets. Many miraculous claims found in the Bible and the Koran-for example, the claims of healings and exorcisms, within the New Testament or the creationist account in the Old Testament-are unreliable. They express the primitive science of an ancient nomadic and agricultural people, and do not withstand contemporary scientific scrutiny.

Unfortunately, some proponents of the historic religions have often used their creeds in order to block or censor scientific inquiry. Freedom of inquiry within science is essential for human civilization; any effort to limit scientific research is counter productive. A good illustration of this is the present effort of some to restrict embryonic stem-cell research on moral or religious grounds. It is argued that if a cell begins to divide, even if only six or eight cells that a "soul" of a person is already implanted and that any effort to experiment with this is "immoral." The postulation of a soul to prohibit scientific inquiry is reminiscent of the suppression of Galileo and the teaching of Darwinism. Thus so far as religion claims to provide some overall imprimatur for scientific research we need *a separation of religion and science.*

The second area concerns the relationship between science and morality. This question can be raised here because many people think that the main function of religion is moral. There is logic of judgments of practice, rules of effective decision-making and ethical knowledge that we can develop quite independently of a religious framework. Science has a role to play here, for it can expand the means at our disposal (technology), and it can modify value judgments in the light of the facts of the case and their consequences. Many people today mistakenly believe that you cannot be moral without religious foundations. Ever since the Renaissance, the secularization of morality has continued quite independently of religious commandments.

A third area, which has been hotly debated in the modern world, is the relationship between religion and the state. Most of the democrats today defend the separation of religion and the state; they say that, though religionists have every right to express their point of view in the public square, religion should be primarily a private matter. Religions should not seek to impose their fundamental moral principles on the entire society. Democratic states should be neutral about professing religious principles.

What then is the proper domain of religion? Is anything left to religion? The final answer could be affirmative. This may surprise skeptics, but it can be concluded that religion and science are compatible, depending of course on what is meant by religion. Religion has performed an important function that cannot be simply dismissed. Religions will continue with us in the foreseeable future and will not easily wither away. Certainly within the political domain, religionists do not have any special competence, similarly for the moral domain. It should be left to every citizen of a democracy to express his or her political views. Likewise, for the developing moral personality who is able to render moral judgments.

If this is the case, the great challenge to scientific naturalism is not in the area of truth but of hope, not of the good but of promise, not of the just but of expectation-in the light of the tragic character of the human condition. This is in contrast with the findings of neo-Darwinism, which recognizes that death is final, not simply the death of each individual but the possible extinction some day in the remote future of the human species itself. Evolutionists have discovered that millions of species have become extinct. Does not the same fate await the human species? Cosmological scientists indicate that at some point it seems likely that our Sun will cool down, indeed, looking into the future, that a Big Crunch may overtake the entire universe. Others talk about a deep freeze. Some star trekkers are inspired by science fiction. They say that one day perhaps we will leave the earth and inhabit other planets and galaxies. Nonetheless at some point the death not only of the individual, but also of our species, our planet and solar system seems likely.

What does this portend for the ultimate human condition? We live in an epoch where the dimensions of the universe have expanded enormously on the micro and on the macro level. We are talking about billions of light years in dimension. Much of this is based on speculative extrapolation. Nonetheless, we can ask, does the naturalistic picture crush human aspiration? Does it destroy and undermine hope? Does it provide sufficient consolation for the human spirit? From this perspective, the central issue for humans is the question of human courage. Can we live a full life in the face of ultimate human extinction? These are large-scale questions, yet they are central for our religious consciousness. Can scientific naturalism, insofar as it undermines theism, provide an alternative dramatic, poetic rendering of the human condition, offering hope, and promise? Countless numbers of brave individuals can live significant lives and even thrive accepting the possible far death of the species and our solar system. But so many other humans-perhaps the bulk of humankind-cannot bear that thought. They crave immortality and religion satisfies their need. Many others do not stay nights worrying what will happen five, ten, or fifteen billion years from now. They find life worthwhile for its own sake here and now.

Wrestling with Mystery

The mystery of life is not a problem to be solved but a reality to be experienced. Aart Van Der Leeuw. When we look at the beautiful Mother Nature, we find therein a flawless incredible beauty. The enchanting rainbow, vast variety of living beings, flora and fauna, gradual development of complexities all seem to be very much simple and at the same time mysterious as well. Where did actually life on this earth come from? How could DNA, RNA and brain develop to this present stage? What happens to the soul when someone dies? Is there a meaning to human life, unrealized by most people? What is our origin? What is our destination? What is our purpose? Seemingly simple questions might have big and intricate answers and in most of the cases we may not find an answer at all.

Death and life are a mystery, and genetic engineering/care of the dying raise profound spiritual questions about human existence. During all recorded human history, there has never been a substantiated case of a living thing being produced from anything other than another living thing. As yet, Evolutionism has not produced a scientifically credible explanation for the origin of such immense complexities as DNA, the human brain, and many other complex elements of the cosmos.

Death constitutes the last chapter of the history of our human life. In many cases the understanding of death penetrates the whole life and is the red thread throughout the activities and volition of human beings. Death is a unique episode at the end of the life of man, and as such it is the object of important studies by the philosopher, the scientist, and the ordinary man. Death is a mystery, and only in the light of everlasting life, has its dreadful threat been transformed into a happy and victorious event for the believer. It is proper to examine the question of everlasting life in the light of the unfailing sources from which we derive our information and establish our belief in this truth. The belief in the existence of God is fortified by the classical proof provided by the existence of the universe and its purpose.

It is highly premature for materialists to claim that all living things evolved into existence, when science has yet to discover how even one protein molecule could actually have come into existence by natural processes. There is no scientific proof that life did (or ever could) evolve into existence from non-living matter. Further, there is substantial evidence that spontaneous generation is impossible. Only DNA is known to produce DNA. No chemical interaction of molecules has even come close to producing this ultra-complex code which is so essential to all known life.

It is even more interesting and profitable to become aware of the reasons for which many scientists, past and present, see rigorously conducted scientific research as not only compatible, but even happily capable of integration with the sincere and joyous recognition of the existence of God. From the considerations which often accompany their scientific endeavours in the manner of a spiritual diary, it is easy to see the intersection of two elements. The first is the way in which research itself, be it great or small, carried out with extreme rigour, always leaves an opening for further questions in an endless process which reveals in reality an immensity, a harmony, a finality which is not explainable in terms of causality or through scientific resources alone. To this is added the irrepressible question of meaning, of higher rationality -indeed, of something or of someone capable of satisfying interior needs- which refined scientific progress itself, far from suppressing, intensifies.

How did life originate? Perhaps no question has mystified mankind more. Did life on Earth-with all its incredible complexity and diversity-arise from an undirected evolutionary process, as many scientists have believed since the time of Charles Darwin? Or was something else at work? Until recently, Darwin's theory has gone relatively unchallenged in the scientific community. But now, powerful new evidence points to an exciting explanation known as the theory of intelligent design. This is a fascinating journey into the interior of a single living cell-an elaborate microscopic world with distinct systems that point dramatically to overwhelming evidence of intelligent design. Featuring state-of-the-art computer animation and persuasive insights from leading scientists, this riveting video examines an idea with the power to revolutionize our understanding of life and to unlock the mystery of its origins.

There are many areas where one can find radically different and confusing explanations. Some of them are:

- Does the soul or consciousness exist, as a separate and distinct entity; or is it a function of the brain?
- Does science provide evidence for "intelligent design" or does evolutionary biology suffice without it?
- Is it possible to influence the healing of persons by praying for them at a distance, or are the tests performed completely unreliable?
- Is there empirical evidence for the claim that "near-death experiences" enable us to reach "the other side" or are there alternative physiological and psychological explanations for these experiences?
- Can mediums under certain conditions communicate with deceased persons, or are the protocols of such tests too loose?
- Does the Big Bang hypothesis in astronomy point to God as the cause of the universe, or is the latter claim beyond science and purely speculative?
- It is said that God created all things. So who created the evil angel? Could a God be deemed good and just, be the creator of misfortune?

In dealing with the above topics various questions emerge: Are coherent theories and testable hypotheses presented? If so, what is the evidence for them? Do paranormal- spiritual-religious explanations survive critical scrutiny? Faith has long been a puzzle for science, and it's no surprise why. By definition, faith demands belief without a need for supporting evidence, a concept that could not be more opposed to the principles of scientific inquiry. In the eyes of the scientist, an absence of evidence reduces belief to a hunch. It places the assumptions at the heart of many religions on the rockiest of ground. So why do so many people believe? And why has belief proved as resilient as scientific progress unravels the mysteries of plagues, floods, earthquakes and our understanding of the universe? By injecting nuns with radioactive chemicals, by scanning the brains of people with epilepsy and studying naughty children, scientists are now working out why. When the evidence is pieced together, it seems that evolution prepared what society later molded a brain to believe.

Interdependence of Science and Religion in solving the Mysteries

One factor in the development of religious belief was the rapid expansion of our brains as we emerged as a species. As the frontal and temporal lobes grew larger, our ability to extrapolate into the future and form memories developed. When this happened, we acquired some very new and dramatic cognitive skills. For example, we could see a dead body and see ourselves in that position one day. We could think 'That's going to be me.' That awareness of impending death prompted questions: why are we here? What happens when we die? Answers are always needed.

As well as providing succour for those troubled by the existential dilemma, religion, or at least a primitive spirituality, would have played another important role as human societies developed. By providing contexts for a moral code, religious beliefs encouraged bonding within groups, which in turn bolstered the group's chances of survival. Some believe that religion was so successful in improving group survival that a tendency to believe was positively selected for in our evolutionary history. Others maintain that religious belief is too modern to have made any difference. While some continue to tease out the reasons for the emergence of religion and its persistent appeal, others are delving into the neuroscience of belief in the hope of finding a biological basis for religious experience. As a starting point, many studies focused on people with particular neural conditions that made them prone to experiences so intense, they considered them to be visions of God.

Russell notes that science also "raises questions which are broader than science can answer." These questions are usually philosophical or moral in nature: Why is there a universe at all? Why do we find mathematical order in the world around us? Why does science itself make sense? Why is science of value? And how can we best put our scientific knowledge to use? According to Russell these questions require the input not only of scientists, but also of philosophers, ethicists, and theologians.

Albert Einstein opened our eyes to the idea that space and time are not fixed and absolute, but elastic and mutable. Einstein sought to comprehend what he called the order "deeply hidden behind everything" and to describe it mathematically. He often spoke of this—half-seriously, half-whimsically—as his longing to understand "what God was thinking." Science is the truth only in matters that can be objectified; in the spiritual world, where values, goals, authority and purpose are located, science has nothing to say. It is a poor life that is restricted to the scientific standard of truth, where we all are nothing but a collection of atoms without meaning and purpose. Realizing the narrow-minded nature of science opens the gate to an understanding of God that complements the scientific truth and gives life, love and peace.

Conclusion

Faith is a personal belief or trust in something. The bases for faith and trust are not always needed to exist with materialistic appearance. Science and religion both can be separately or collaboratively adapted to test faith but none of them assures us of a certain and flawless result. It is quite possible that in spite of much scientific research and religious thinking we remain where we started from on the journey of testing faith. Further, after much of useless efforts, as it comes out at the moment, we must learn to live with mystery. We are not smart enough; we do not see enough, we do not understand enough. None of our vaunted technological equipment will answer these questions. That is one of the most difficult lessons to learn in life. We think that because God tells us certain things about himself we can figure out what he is going to do. We must resist that; we cannot God will never be false to his character; he will never contradict what he said. We are just not smart enough to always figure it out or anticipate it. Life is a mystery and we need a strategy to try to solve the mystery of life, and if we do not solve this mystery to it's fullest then also life remains beautiful and assuring as it has always been. Science and religion both can be used for solving mysteries but an intelligent brain should always try to keep a balance between the extremes offered by these two life-governing agencies. The best way to deal with mysteries could be to live life with a faith–faith in nature, faith in life itself and faith in the Almighty.

Notes

1. http://nowscape.com/mormon/faith_def.htm
2. http://www.thefreedictionary.com/faith
3. Satellites experience measurable time dilation, buses do not; schoolchildren are still taught that Every Action has an Equal and Opposite Reaction, despite quantum mechanics.

References

http://nowscape.com/mormon/faith_def.htm

http://www.thefreedictionary.com/faith

Gould, Stephen J. (1999), "Non-Overlapping Magisteria." *Skeptical Inquirer* July/August 23 (4).

Quantum Cosmology and the Laws of Nature: Scientific Perspectives on Divine Action (CTNS & Vatican Observatory/UNP, 1993)

Chaos and Complexity (CTNS & Vatican Observatory/UNP, 1995)

Kanjirathamkunnel Shijo, CM (2004), *"The Miracle of Double Helix."* Vidyankur, Journal of Philosophical and Theological Studies. Vol. 6, February, 2004.

Job Kozamthadam (2001), "The Genome Revolution and the Science-Religion Dialogue I." Vidyajyoti, Journal of Theological Reflection. Volume 65, No. 2, February 2001.

ECOLOGY–A PYTHAGOREAN PERSPECTIVE

Mita Darbari

The term "ecology" is derived from the Greek word *oikos* meaning "household".[1] The house (*oikos*) is a place where its inhabitants relate to one another and dynamically interact. As the house provides a shelter for people to live, in the same way, the universe provides a haven for the infinite manifold of unanimated and animated forms to exist and evolve together. Pythagoras may have been the first to use the term *cosmos*[2] to imply that the entire universe has order, which could be understood by mathematics as he postulated that the natural laws of harmony of the microcosm of music, of the human soul, biological life and of the cosmos were identical.

Pythagoras, considered as the father of pure mathematics, tried to answer many philosophical questions, which are still uppermost in the minds of those who are concerned with existence, ethics, the human condition, and our relationship to the nature. The philosophy of numbers that Pythagoras is associated with is both mystical and mathematical which incorporates mathematical laws and geometric figures as proof that numbers are fundamental elements of the universe. Pythagoras saw the cosmos as structured according to number in so far as the *tetraktys* is the source of all wisdom. His cosmos was also imbued with a moral significance, which is in accordance with his beliefs about reincarnation and the fate of the soul. The main thrust of Pythagoreanism is that men are already divine; we achieve immortality not by becoming divine, but by realizing that which is already within us, a concept of contemporary deep ecologists.

One, the Monad, was the principle of Unity from which all things begin. Two, the Dyad, was Duality. It was both the beginning of strife and the possibility for the development of relationships between things, as everything is no longer the same. Pythagorean dualism is expressed in terms of ten contrarieties (The Ten Pythagorean Principles): limited and unlimited, odd and even, one and plurality, right and left, male and female, resting and moving, straight and curved, light and darkness, good and bad and square and oblong. The Triad, Three, forms a bridge, allows relation between the two extremes. The reconciliation of these opposites creates harmony. For human behavior, this harmony resulted from behaving temperately or by exhibiting moderation.

Pythagoras taught that both man and the universe were made in the image of God; that both being made in the same image, the understanding of one predicated the knowledge of the other. He further taught that there was a constant interplay between the Grand Man (the universe) and man (the little universe). Pythagoras said that man was a microcosm; which means, a compendium of the universe not because, like other animals, even the least, he is constituted by the four elements, but because he contains all the powers of the world. For the world contains Gods, the four elements, animals and plants. All of these powers are contained in man. He has reason, which is a divine power; he has the nature of the elements, the powers of moving, growing, and reproduction. In this way the universal principles extend to man.

However, in each of these he is inferior to the others. Than the gods, we have less reasoning powers; and less of each of the elements than the elements themselves. Our anger and desire are inferior to these passions in the irrational animals; while our powers of nutrition and growth are inferior to that in plants. Constituted therefore of different powers, we have a difficult life to lead.[3]

Although dualism was evident throughout Nature, the fusion of opposites held the secret of balance. Only in Oneness was true identification with God possible. God was equated with love, which pervaded the entire universe. The soul of the individual and that of the Earth and universe were intimately linked. This macrocosm idea, which seems ever more pertinent today in a world racked by fragmentation, is similar to the comparative recent concept of the absolute relativity of existence. But to Pythagoras, this was not merely confined to the material world. He viewed death as a passageway on the continuing journey of the immortal soul, whose re-incarnation would be determined by the values of previous deeds. A devout vegetarian, who considered animal sacrifices in the temples to be a sacrilege, he believed that pain inflicted upon Nature would inevitably return to haunt humankind, both spiritually and physically.

Pythagoras taught that the soul was immortal and that after death it transmigrated into other animated bodies. After certain specified periods, the same events occur again; that nothing was entirely new. And hence Pythagoras taught that friendship was the truest and nearest perfect of all relationships. He declared that in Nature there was a friendship of all for all; of gods for men; of doctrines one for another; of the soul for the body; of the rational part for the irrational part; of philosophy for its theory; of men for one another; of countrymen for one another; that friendship also existed between strangers, between a man and his wife, his children, and his servants. Pythagoras taught that all animated beings were kin, and should be considered as belonging to one great family and so he believed in universal friendship and hence he commanded:

> *Not to destroy or injure a cultivated tree, nor any animal either which does not injure men.*[4] The morality of Pythagorean ethics encouraged the creation of all-encompassing laws to make the killing of living creatures, animal sacrifices, blood sports and, of course, the eating of meat, illegal. Pythagoras not only saw vegetarianism as the key to peaceful human co-existence but also recognized the health advantages of a meat-free diet. ... *he even forbade men to kill animals at all, much less would have allowed his disciples to eat then, as having a right to live in common with mankind.*[5]

Furthermore, he did not worship at altars where animals were sacrificed:... And that he used to forbid them to offer victims to the Gods, ordering them to worship only at those altars which were unstained with blood.[6] The important discovery that the relations between the notes of musical scales can be expressed by means of numbers is a Pythagorean one. Pythagoras reasoned if one rolls a ball on the earth, it would make a noise. Therefore, a large ball, such as the moon, or sun, or the planets cannot fail to make a large noise. The planets each produce a musical tone, and their orbits were laid out in proportion like the notes in a musical scale. Together they create a Celestial Harmony. Pythagoras conceived of three kinds of music: 1. Musica Instrumentalis or music made by people playing instruments or singing. 2. Musica Humana or the music which human beings are, e.g. he likened people's bodies to the body of a guitar, and our nervous systems to the strings. 3. Musica Mundana, or the Music of the Spheres.

And the trouble with people is that the music, which they are (Musica Humana), gets out of harmony with the Music of the Spheres. This is the root cause of disease, anger, depression, and anxiety; in short all human ills. Therefore he composed musical remedies, tunes that affected the human organism (Musica Humana) in such a way as to bring it back into harmony with the Music of the Spheres.[7]

For Pythagoreans, all souls spring from the same source and inhabit every living thing. It follows from this that all living things are related and there is literally a kinship of nature. Without knowing, the Pythagoreans were propounding a movement about two thousand and five hundred years ago, which promotes an awareness of the oneness and interconnection of all life and its cycles of change and transformation i.e., deep ecology.

Their beliefs about the soul in conjunction with their view the Cosmos is harmonious leads them to the ethical position that the task for human beings is to ensure that they live in conformity with the harmony of the cosmos. This can be achieved by following the ethical principles laid down by the community. Concluding from Divine Harmony: For Pythagoras, mathematics served as a bridge between the visible and invisible worlds. He pursued the discipline of mathematics not only as a way of understanding and manipulating nature, but also as a means of turning the mind away from the physical world, which he held to be transitory and unreal, and leading it to the contemplation of eternal and truly existing things that never vary. He taught his students that by focusing on the elements of mathematics, they could calm and purify the mind, and ultimately, through disciplined effort, experience true happiness.[8]

Notes and References

1. Encyclopedia-msn.Encarta. Online. Cited on 13-02-2006. http://encarta.msn.com/encylopedia_761576703_4/Ecology.html#howtocite
2. Diogenes Laertius, *Vitae philosophorum*; XXV.
3. *Anonymous Biography of Pythagoras*; 15; Preserved by Photius. Online. Cited on 07-02-2006. http://www.completepythagoras.net/volume1/photus.html.
4. Diogenes Laertius, *Vitae philosophorum*; XIX.
5. Diogenes Laertius, *Vitae Philosophorum*, XII.
6. Diogenes Laertius; XIX.
7. Hall, Manly P.; *Secret Teachings of All Ages: The Pythagorean Theory of Music and Colour*; by H.S. Crocker Company, incorporated MCMXXVIII; San Fransisco; 1928. Online. Cited on 14-08-2005. http://www.sacred_texts.com/eso/sta19.htm.
8. John Strohmeier and Peter Westbrook. *Divine Harmony: The Life and Teachings of Pythagoras*. (Berkeley, CA: Berkeley Hills Books, 1999), p. 66.

MAN, NATURE AND CULTURE-REDEFINING ECOLOGICAL BALANCE

Elena Philip

The basic assumption in the theory of Elements is that, like the rest of the material world man is made up of Elements which at death disintegrate and dissolve into nature. To make the physical man capable of transforming into a cultural person correlations are established between the natural and the supernatural orders at all levels of human existence. A bio-spiritual principle operates. Cultural activities are determined by the body, mind and spirit. Satisfaction of the bodily needs is a critical factor in the survival of man. The mind responds to the physical needs-it creates the urge to satisfy these needs. The spirit which transcends both, the body and the mind, mediates between and allows the satisfaction of the needs to such an extent that the world order is not disturbed. Man is not the measure of all that exists. He can neither create nor destroy nature. He is a creator of culture and that also only in a limited sense, because cultural forms are prone to the five fold order of transcendence.

Human culture is unique only in that it allows humans to satisfy their impulses in a far more complex manner than animals. Another peculiarity which distinguishes man from animals is his innate capacity and will to transcend even the physical needs. This he often does for spiritual gains and emotional satisfaction. The urge for transcendence is attributed to the complex structuring of the sense-organs, and all that is known as elements. Man is related to nature, the elements and animal and plant life. The environment in which he lives is not an alien environment. He always considers it his own, where he is like all other beings, but endowed with the special gift of self-reflection and speech. Indeed, man is constantly seen as an embodiment of the elements and forces of nature and in relationship to animal and plant life. This gives the world a different character from what is implied in the modern idea of progressive evolution. Man is not the best because he overpowers and conquers nature and is thus the fittest to survive, but he is one amongst the many with the capacity for 'consciousness' and self-reflection and transcendence from his pure physicality, through psychical discipline.

Today, in science the basic oneness of the universe is clear. It has given rise to many unified field theories, suggesting that the constituents of matter are interconnected, interrelated and interdependent. This basic phenomenon may be understood not in terms of isolated entities, but as integrated parts of the whole. Without going into each theory, and not being competent to do so, one may yet say that the physical universe is proving to be a seamless texture of inseparable events and entities, organized in accordance with a universal principle that specifies itself in innumerable forms.

In the last few decades man has, through his power to transform his environment, wrought accelerated changes in the balance of nature. The result is frequent exposure of living species to dangers which may prove irreversible. Despite the traditional holistic view embedded in each and every discipline, articulated through sacred incantations and systematized as ritual for constant reminder of the need to sustain and foster the ecological balances of nature, we stand at the threshold of disaster in a manner never before faced by man, who as the Tbilisi declaration states, has "used his power to transform his environment". We may even substitute the word 'transform' with words like 'conquer', 'destroy', 'desecrate', 'plunder', so as to bring home the fact that a single factor of Man who can be distinguished from all living matter through his distinctive power to reflect, to articulate and be wise, has been the very instrument of these ecological imbalances - imbalances which threaten the very existence of Man. As one eminent scientist - philosopher put it in the context of India: "Even more than population explosion, the imbalance in the environment and ecology is the greatest threat to the continent. We may be able to control the first, but the process of devastation we have begun in the second may bring final doom."

The evolution of mankind followed biological, social, and cultural pathways, presumably in this sequence. In the evolutionary sense life is maintained as a result of the genetic transmission of the structures determining the needs and functions, whereas humanity is maintained through the cultural heritage (transmission of verified experiences) deciding upon the existence of a society, in particular, under given conditions (to which it is not adapted biologically). The humans adapt to the environment almost directly through their culture. There are, however, many ways of survival, depending on the conditions under which a given culture is formed. Culture, is subject to evolution which is of adaptive importance in the face of the new environmental conditions and has conservative importance to the species and social values of the earlier generations.

Change can be progressive. Progress can also be called modernization. It is not univocal which elements of culture are old-fashioned and which are modern, or which are regressive and which are progressive. Each time this requires a separate evaluation with respect to a certain criterion. This criterion can be of adaptive value. Modern civilization is associated with the elements of urbanization and industrialization processes. It is thus worth considering the consequences and cost of modernization to the environs of humans. A question arises whether modernization always means progress. The intended favorable changes have different side effects and because of their high rate and large scale, one may ask if the total effects can be called progress. This is an important ecological problem.

Modernization should be associated with the comfort of life. But this is a controversial issue since for most people a comfortable life is associated with material goods, and the side-effects of their production, utilization, and disposal typically account for degradation of the environment and deterioration of well-being. It is important not to oppose modernization to tradition. Modernization rooted in tradition should contain adaptive values, determining persistence and worthiness of life. The humanization process occurred over millions of years ago. Generations of nomadic groups of humans extended their range over distinct climatic regions and man himself changed accordingly. Nomadic groups settled because they started to cultivate the land, and hence permanent human settlements developed. These settlements became the basis of urban civilization. Genetic adaptations occurred through changes between generations. A generation is therefore the unit of biological evolution.

Civilizations are typically defined as very advanced forms of social existence, based on the written language, arts, belief (religions), and relatively high technology. They are cultural expressions of a specific form of socio-political organization adequate to the environment. Civilizations emerged, developed, and died, like species in nature. A universal trait of mankind that developed in the process of evolution, as a non-biological form of satisfying the needs along with the preservation of the common interest is culture. The essence of biological existence lies in the struggle for existence between organisms, whereas the essence of the social existence of the humans lies in the altruism understood as humaneness.

Modernization is a new kind of adaptation. A new culture represents a new kind of linkage with the environment. Modernization is a cultural adequacy to the actual environmental conditions, including natural, social, and technical surroundings. New needs cause actions forming new cultural environment. In the evolutionary perspective they lead to the transformation of civilization. Progress thus involves everything of positive value to mankind. Other changes cannot be called progress from the point of view of human evolution.

The spiral of progress works in the following way: humans adjust to new conditions physiologically, and psychologically they are habituated to them soon. If these conditions are favorable, the satisfaction with them gradually disappears. A need for new conditions satisfying the requirements and aspirations develops, releasing the initiative of new changes. But in these new conditions, the feeling of satisfaction vanishes again with time, giving rise to new needs and aspirations. An increase in the consumption of goods combined with environmental pollution can have a deteriorating net effect on the quality of life. This happens because side-effects emerge with a delay, they are often difficult to predict, and do not always impress the imagination.

As a result of the specific character of human nature, a change of needs takes place at certain time intervals or permanently. This has an effect on the innovator activities and civilization progress. Modernization conceived in this way leads to hygienic activity, increased safety of life, easier or more efficient work. This change in the mode of life and living conditions accounts for a decline in mortality and longer life span, directly leading to a higher density of the population, and indirectly to a numerical increase in the needs. The last two factors enhance both consumption and production. When the production and consumption exceed a certain threshold, they account for the exhaustion of non-renewable resources, environmental pollution, and stress.

The consequences of changes in the mode of life and the environment are most seen in the periods of the highest entropy in human development The unbridled consumption related with wastefulness of goods, which now-a-days threatens in various forms of the pollution and exhaustion of non-renewable resources. An example of this direction is the excessive production of goods and the philosophy of life in which "to be" is replaced by "to have". And here moderation should really be recommended. Modernization is a new outlook on the world, where intelligence determines the ability to detect patterns and consumption defines the ability to utilize these patterns. They are the different aspects of modernization: cognition and action. The value of these two aspects can be measured by the effectiveness of the adjustment to the environment. The values of modernization range from positive to negative and real values of the same changes can be different, depending on the situation, eg. environmental conditions and habits which are typical of different civilizations.

The disturbance in water, earth, vegetation, river and mountain ecology systems has naturally threatened all manner of life - aquatic, terrestrial and celestial worst of all, there is the pollution of the holy space - the air, and the sky which permeates and envelopes all life. Destruction is writ large on the balmy skies, once azure blue, today smoke grey. Water, the basic sustaining principle - clean water or to use a traditional phrase 'pure or unpolluted -water' - is becoming scarce. Water systems are increasingly over-exploited and polluted. Arid lands have increased, soil which was venerated has been eroded and infertility, sand and salt have taken over.

Related to the pollution of water and earth is the massive unprecedented deforestation: the soil's ability to absorb and hold water has diminished. Severe floods have occurred more frequently and deforestation has affected most adversely the eco-system. Each day we learn of the increase in wastelands and the consequent effect upon rural life. Desecration of the bowels of the earth through excessive quarrying is common.

The rapid degradation of the natural environment have been irrevocably disturbed with mass scale deforestation, desertification, chemical and nuclear pollution. The earth, oceans and rivers are increasingly poisoned. Global warming, with encroaching drought, is threatening large areas of habitable lands. The windswept effects of pollution and radiation spreading from one corner of the planet are affecting all others while the very ether itself is being radically rearranged as is becoming evident in the ozone holes. At the same time large oxygen producing areas are dangerously being deforested.

Therefore man aspires higher and higher, beyond the pollution of his making, to that one source which sustains all and is yet beyond his reach, namely, the Sun-the giver of light in the sky and fire on earth-Agni-the source of energy which appears unpollutable unlike the fuel of his making. This energy, physically and metaphysically, is man's only ray of hope for the continuance of life. Perhaps these will suffice to convince us that stated differently or stated in the traditional language understood by the literate and illiterate of this country, Man has disturbed the cosmic order, the rhythm of the movement of the earth, water, fire, air and space , i.e., the five elements, where interaction, inter-connection and inter-weaving was the rule rather than the exception. It must be remembered that no single element is autonomous. It is in their ecological interaction that they assume significance.

Cutting across historical developments, philosophic debates, scientific controversies, religious sects and cults, the one principle which underlies and provides unity as also continuity of vision and perception is the assertion that Man is only one among all living matter. Man's life depends upon and is conditioned by all that surrounds him and sustains him, namely, inanimate, mineral and animate, aquatic, vegetative, animal and gaseous life. It is, therefore, man's duty constantly to remind him - in individual and collective fear; it is wisdom contained in the language of myth and symbol. Their efficacy lays in their capacity of multiple interpretations at the biological function, societal, philosophic and religious levels.

We must understand that the whole Earth is one interconnected entity. Everything upon this earth has intrinsic value - a tree, a worm, a river, all and everything are good in themselves. The most important thing is that everything upon the earth has a deep intrinsic value; all things maintain a deep intrinsic relationship with each other. They are good in themselves.

We have no right to think that we human beings are more important. Everything that naturally exists has its own natural balance and harmony.

Once we accept that the earth is good, how do we interact with it? We human beings need food; we have to cultivate land; and we have to fulfill our vital needs. When we are tilling the soil, or making a product, we need to do it in a sustainable way. Whatever one does, one must remember how their action is going to affect the coming generation. We must think of posterity, and how they are going to be affected by what we do today. We need to design all our activities in such a way that all designs for living contain the idea of permanence. We need to bear in mind the question, whether it is sustainable or not? Is it only for short-term profit, or is it a long-term, continuous and durable design? The modern industrial society does not have a sense of reverence for nature, and this results in the pollution and degradation of the earth. The crisis of environment comes out of a utilitarian, materialistic, non-sacred, non-spiritual world-view- "the Earth is there for us to use, for our comfort, for our convenience." As a consequence we have taken from nature without knowing its limits. When we have a sense of reverence, we shall take from nature only what meets our vital needs.

Mahatma Gandhi said, "There is enough for everybody's need in this world, but not enough for anybody's greed." So the need and greed have to be differentiated. How can you differentiate them? A government cannot make laws for it or cannot force it. It has to emerge out of our own individual heart, from a sense of beauty, a sense of the divine. When we have that, then we take things from the earth and try to replenish her for what we have taken. The earth not only provides enough for humans, animals and birds to eat, but also enough to return back to the earth; the peels, the straw, the skin, the fruits and vegetables have plenty of good for us to eat and plenty to put back into the compost which goes back into the earth. Thus the earth is replenished. Nature is our great teacher, and we can learn to replenish and not waste. There is no greater teacher than nature. Spirituality is not dependent on any organized religion. It is a sense in your heart that there is much more to life than meets the eye; there is a greater mystery than we can know or measure; and there is greater meaning behind the world of appearance. The light is burning inside us. We need to close our eyes and look within. The spiritual light is inside our soul.

The world cannot be saved just by the technocrats, or by ecologists. They say, "We can manage the environment." But everyone knows that environment cannot be managed. We can only revere environment; we can only respect environment; and we can only see environment as part of us and us as a part of environment. This total unity can come only when we have a spiritual base and not just a utilitarian base. In most ancient or 'traditional' societies a deep respect for the elemental forces in which human beings found themselves immersed, is evident from the cultural histories, rites and artifacts that have been passed down to us. In the history of humankind, all cultures are faced with the same dangerous dilemma of a serious disregard for the natural order of the elements upon which existence is dependent.

Perhaps it is for the first time in history that the various cultures have access to each other's traditions and histories to be able to develop a healthy respect through understanding, if not an interdependent trans cultural foundation for a harmonious relationship with the earth and its beings. Today the greatest danger to human survival is humankind itself. Without checking human greed and intolerance no amount of environmental friendly actions will ensure a future for man. In this light cultural sensitivity, with understanding, play the most vital role in transforming and maintaining a harmonious relationship with our physical, social and spiritual environment.

The very fact that we now talk about global environment, which during the last two centuries of global industrialization we have somehow ignored in the blind pursuit of 'progress', shows how far the world's cultural systems have gone astray. We can no longer point a finger at one group or in one direction, for in the globally interdependent ecology there is no east or west, nor north or south. Industrialization and human addiction to greed has become a world-wide phenomenon. What we can, and must do, if we are to survive as a species, is develop a culture based on an awareness of our physical and psycho-physical situation. To be effective we need to bring the awareness and sensitivity of these basic underlying principles to the forefront of the human community which will require a dynamic and energetic reviewing of the process and necessity of a cultural ecology. This enormous task will require national and international network of the cultural communities getting to know one another and collectively exploring the ecological imbalances and finding effective ways to transform human relationship with mother earth.

> Pure and peaceful be earth,
> Peaceful ether, peaceful heaven,
> Peaceful water, peaceful herbs, peaceful trees,
> May all gods and environs be pure and peaceful;
> May there be purity, non-pollution and peace through these invocations.

It has become essential to collectively develop a respectful understanding for the world in which we find ourselves. To do this we can learn much from the ancient societies who had developed cultural attitudes and practices which were in relative harmony with the elemental forces of both nature and human nature. Therefore, the lessons are obvious. Non-pollution, discipline, restraint, awareness of inter-dependability and inter-relatedness is taught to us through custom, daily routine, myth and ritual, but we don't learn. We must learn before it is too late.

References

Malville, McKim John, *Cosmogony and the Elements - The Intuition of Cosmos in Science and Myth.* www ignca.nic.in

Vatsyayan, Kapila, *Ecology and Indian Myth* . Vedamsbooks. Sagepub.com.

Saraswati Biadyanath, *Elements of Nature and the Order of Culture.* Indira Gandhi National Centre for Arts Central Vista Mess, Janpath, New Delhi.

Saraswati Biadyanath, *Lifestyles in traditional cultures: A conceptual framework.* ignca.nic.in

Adhikary, A.K. (1992), "The Birhors and their universe". Seminar Paper. New Delhi: Indira Gandhi National Centre for the Arts.

Malik. S.C. *Holistic Science and Consciousness,* Indira Gandhi National Centre for the Arts,New Delhi.

Kapila Vatsyayan ed. (1995), Wolanski Napoleon. *Modernization as a Form of Cultural Adaptation to the Environment Prakrti,* The Integral Vision.

Futterman, J.A.H. *Science and Faith* www.mat.univie.ac.at

Mulvihill Bryan, *Mahabharata in Determining Cultural Ecology,* Aesthetics determines Cultural Ecology, Vedame Books (P) Ltd. New Delhi.

Singh T.D. *Life and evolution.* Singh T.D. Life and spiritual evolution, by T.D.Singh Bhaktivedanta Institute 2005 Kolkata.

Singh T.D. *Science and Religion.* www.metanexus.net/conference 2005.

Singh T.D. *Vedanta and Science Series.* Bhaktivedanta Institute, Kolkata.

Singh T.D. *Towards a Culture of Harmony and Peace.* Bhaktivedanta Institute, Kolkata.

Savijnanam-*Scientific Exploration for a Spiritual Paradigm,* Vol. 1,2,3 & 4 Prakrti, Sundeep Prakashan, New Delhi, Vol. 1.

Savijnanam, Bhaktivedanta Institute (2002), Sundeep Prakashan, New Delhi.

Savijnanam Bhaktivedanta Institute, 2003: , Sundeep Prakashan, New Delhi, Vol. 2.

Satprem-On the way to Supermanhood-Essay of experimental evolution, Mira Aditi & The Mother's Institute of Research , New Delhi.

Singh, T.D. ed., Even Noble Laureates on Science and Spirituality. Bhaktivedanta Institute. Kolkata.

Science, Spirituality and Nature of Reality. A Discussion between Sir Roger Penrose and T.D. Singh. Bhakti Vedanta Institute, 2005. Kolkata: www.iisc.ernet.in

SCIENTIFIC PERSPECTIVE OF THEOLOGY: A STUDY OF FAITH AND SCIENCE

Vidhu G. Singh & Ashish S. Noel

Faith does not offer resources to scientific research as such, but it encourages the scientists to pursue his research knowing that he meets, in nature, the presence of the creator. Some of us are walking along this way. All of us are concentrating on our intellect. There exists a rather widespread notion that men of science are generally agnostics and that science leads one away from God. The extraordinary advances of science, particularly over the last two centuries, have sometimes led to the belief that it is capable of answering by itself all of man's questions and of resolving all his problems. Some have concluded that by now there is no longer any need for God. For them, faith in science has supplanted faith in God. It has been said that one must choose between faith and science: either one embraces one or believes in the other. He who proceeds with a commitment to scientific research no longer has need of God; vice versa, he who wishes to believe in God cannot be a serious scientist, because between science and faith there is an irremediable conflict.

If methodical investigation within every branch of learning is carried out in accord with moral norms, it never truly conflicts with faith. For earthly matters and the concerns of faith derive from the same God. Indeed whoever labours to penetrate the secrets of reality with a humble and steady mind is, even unawares, being led by the hand of God, who holds all things in existence, and gives them their identity. It can in fact be pointed out that there have always been, and still are today, men of science who in the context of their human scientific experience have positively and beneficially believed in God. It is even more interesting and profitable to become aware of the reasons for which many scientists, past and present, see rigorously conducted scientific research as not only compatible, but even happily capable of integration with the sincere and joyous recognition of the existence of God.

From the considerations which often accompany their scientific endeavors in the manner of a spiritual diary, it is easy to see the intersection of two elements: the first is the way in which research itself, be it great or small, carried out with extreme rigor, always leaves an opening for further questions in an endless process which reveals in reality an immensity, a harmony, a finality which is not explainable in terms of causality or through scientific resources alone. To this is added the irrepressible question of meaning of higher rationality-indeed, of something or of someone capable of satisfying interior needs-which refined scientific progress itself, far from suppressing intensifies.

It is true that the step to religious affirmation is not achieved per se by virtue of the experimental scientific method, but rather by virtue of elementary philosophical principles such as causality, finality, sufficient reason, which a scientist, as a man, finds himself exercising in his daily contact with life and with the reality he studies. Indeed, the scientist's condition as a sentinel in the modern world, as one who is the first to glimpse the enormous complexity together with the marvelous harmony of reality, makes him a privileged witness of the plausibility of religion, a man capable of showing how the admission of transcendence, far from harming the autonomy and the ends of research, rather stimulates it to continually surpass itself in an experience of self-transcendence which reveals the human mystery.

Then if we consider that today the broadened horizons of research, especially in what concerns the very origins of life, pose troubling questions regarding the right use of scientific conquests, we are not surprised by the increasingly frequent request on the part of scientists for sure moral criteria capable of freeing man from arbitrary willfulness. And who if not God, is able to establish a moral order in which the dignity of man, of every man, is firmly cared for and promoted?

Certainly the Christian religion, while it cannot consider certain professions of atheism or agnosticism in the name of science as rational, is equally firm in not accepting affirmations regarding God, which arise from tendencies that are not rigorously attentive to rational processes. At this point it would be very beautiful to make heard in some way the reasons for which not a few scientists positively affirm the existence of God, and to see by what personal relationship with God, with man, and with the great problems and supreme values of life they are sustained. How often silence, meditation, creative imagination, serene detachment from material things, the social significance of discovery and purity of heart are factors which open to them a world of meaning which cannot be disregarded by anyone who proceeds with equal faithfulness and love towards the truth.

When a young person is told: Look, there is a new star, a galaxy, a neutron star 100 million light-years away, yet the protons, electrons, neutrons and mesons which are found there are identical with those which are found in this microphone. Identity excludes probability. That which is identical is not probable. Therefore there is a cause, outside of space, outside of time, the master of being, which made being to be in this way. And this is God.

IMPORTANCE OF PLANTS TO HUMAN IN PHARMACEUTICALS OF 21ST CENTURY

J. Dutta & A. Nandurkar

Introduction

For centuries people have used plant products as therapeutic agents in treatment of many diseases. Plant products–as parts of foods or botanical potions and powders–have been used with varying success to cure and prevent diseases throughout history. Records about medicinal plants date back at least 5000 years to the Sumerians.[1] In 1897, Friedrich Bayer and Co. introduced synthetic acetyl salicylic acid (aspirin) to the world to give the idea of strong historic bond between plants and human health. Aspirin is a safer synthetic analogue of salicylic acid, an active ingredient of willow bark as a remedy for aches and fevers,[2] which replaced natural extracts. The increase in number of pharmaceutical industries had a tremendous impact on disease treatment and prevention. However, the benefits of modern drugs are felt primarily in developed countries, and developing countries continue to rely on ethnobotanical remedies as their primary medicines.[3] The bond between plants and health was felt not only in the area of medicines. Plants are slowly making a comeback in several areas of human health (i.e. functional foods, dietary supplements and recombinant protein manufacturing); they are still losing importance in areas such as the traditional drug discovery process.

Single-ingredient Drugs

Till date plants were an important source for the discovery of novel pharmacologically active compounds, with many drugs being derived[4] directly[5] or indirectly from plants.[6] But during the twentieth century, the emphasis was made on synthesizing these compounds or their analogues synthetically. Now natural products are designed to make perfect new drugs, referred to by industry as New Chemical Entities (NCEs). The contribution of plants to disease treatment and prevention is enormous. In the 1970s, 25% of all drugs dispensed in the USA contained compounds derived from flowering plants,[7] with an even greater proportion of phytochemicals used as drugs worldwide;[8] 16% of drugs dispensed in the USA were derived from microbial and animal sources.[9] Even at the dawn of the twenty-first century, 11% of the 252 drugs considered as basic and essential by the World Health Organization were exclusively of flowering plant origin.[10] The greatest achievement of plant-derived drugs was probably felt in the cancer treatment, where taxol, vinblastine, vincristine and camptothecin have dramatically improved the effectiveness of chemotherapy against some of the deadliest cancers. The most important pharmaceuticals still derived from plants - directly or as precursors - are listed in Table 1. Galantamine and camptothecin are the most recent additions to the plant-derived NCE pharmaceutical pipeline, although both were isolated a long time ago and have spent many years in development. Details are given in table 1.

Table 1

Name	*Source*	*Therapeutic use*
Camptothecin	*Camptotheca acuminata* Decne	Antineoplastic
Cocaine	*Erythroxylum coca* Lamarck	Local anaesthetic
Colchicine	*Colchicum autumnale* L.	Antigout
Galanthamine	*Leucojum aestivum* L.	Cholinesterase inhibitor
Nicotine	*Nicotiana* spp.	Smoking cessationtherapy
Quinine	*Cinchona* spp.	Antimalarial
Tubocurarine	*Chondodendron tomentosum* Ruiz, *Strychnos toxifera* Bentham	Skeletal muscle relaxant
Vinblastine, vincristine	*Catharanthus roseus* L.	Antineoplastic
Artemisinin	*Artemisia annua* L.	Antimalarial
Diosgenin, hecogenin, Stigmasterol	*Dioscorea* spp.	Oral contraceptives and hormonal drugs
Taxol and other taxoids	*Taxus brevifolia* Nutt.	Antineoplastic
Digoxin, digitoxin	*Digitalis* spp.	Cardiotonic
Podophyllotoxin	*Podophyllum peltatum* L.	Antineoplastic

The enthusiasm for using plant extracts for the discovery of novel pharmaceutical leads has declined in the past decade, with many pharmaceutical companies closing or downsizing their natural products groups. Throughout human history plants were unchallenged as sources of new drug discovery but the recent competition from combinatorial[11] chemistry[12] and computational drug design[13] has put an end to the dominance of natural products in drug discovery. It is even possible to suggest that, after thousands of years of bioprospecting for plant-derived pharmaceuticals, people have now identified the most useful, relatively abundant molecules, and that finding new ones would require sophisticated approaches that have not yet emerged. However, many experts believe that the majority of plant-derived natural products–possibly valued at billions of dollars–remain undiscovered or unexplored for their pharmacological[14] activity.[15]

Many living diverse plant species contain a much greater diversity of bioactive compounds than any chemical library made by humans. Few researchers doubt that plants and other sources of natural products are superior sources of molecular diversity and novel molecular chemotypes, particularly in the areas where good synthetic leads do not exist.[16]

The concept of growing crops for health rather than for food or fiber is slowly changing plant biotechnology and medicine. Rediscovery of the connection between plants and health is responsible for launching a new generation of botanical therapeutics that include plant-derived pharmaceuticals, multicomponent botanical drugs, dietary supplements, functional foods and plant-produced recombinant proteins. Many of these products will soon complement conventional pharmaceuticals in the treatment, prevention and diagnosis of diseases, while at the same time adding value to agriculture. Such complementation can be accelerated by developing better tools for the efficient exploration of diverse and mutually interacting arrays of phytochemicals and for the manipulation of the plant's ability to synthesize natural products and complex proteins. This review discusses the history, future, scientific background and regulatory issues related to botanical therapeutics.

So should we abandon plants as sources of NCE-based drugs, embrace them again or redesign strategies for discovering novel molecules from plants? The last approach makes most sense because, despite the remarkable progress made by chemistry, pharmacology, molecular biology, genome research and high-throughput screening, the NCE pipelines of pharmaceutical companies are at historically low levels.[17] Not only is the pharmaceutical industry finding it difficult to replace old products with new and more effective alternatives, it is also having problems developing new products, although many new molecular targets have been discovered recently. A 40% increase in the R&D spending in pharmaceutical research from 1996 to 2001 did not overcome this problem.[18] The lack of reproducibility of activity for >40% of plant extracts[19] is one of the major obstacles in using plants in pharmaceutical discovery, despite the great diversity of compounds they synthesize.

Development of novel technologies that allow rapid isolation and characterization of putative lead molecules and new screening methods more compatible with complex mixtures will be imperative if plants are to return to the mainstream of drug discovery efforts. These technologies should consider and possibly use the fact that the biological activity of plant extracts often results from additive or synergistic effects of its components. Another strategy is to exploit the qualitative and quantitative variations in the content of bioactive phytochemicals, which are currently considered major detriments in phytochemical NCE discovery. Different stresses, locations, climates, microenvironments and physical and chemical stimuli, often called elicitors, qualitatively and quantitatively alter the content of bioactive secondary metabolites. Enzymatic pathways leading to the synthesis of these phytochemicals are highly inducible[20]. This is particularly true for phytochemicals that are well documented for their pharmacological activity, such as alkaloids,[21] phenylpropanoids[22] and terpenoids[23] whose levels[24] often increase by two to three orders of magnitude following stress[25] or elicitation.[26]

The US Federal Food and Drug Administration (FDA) recently published guidelines for standardized multifunctional and multicomponent plant extracts, botanical drugs, under the New Drug Application (NDA) Approval Process.[27] Because of public demand for trustworthy and effective alternatives to NCE pharmaceuticals, the agency proposed abbreviated preclinical and clinical testing protocols for botanical drugs derived from plants. This has enabled US industrial and academic scientists to become involved in botanical drug R&D efforts. Botanical drugs are fully accepted and widely prescribed in China, Japan, India and other Asian and African countries. In addition, some countries in Europe, such as Germany, allow physicians to prescribe botanical drugs.

The NCEs were used but recently attempts have been made to treat complex diseases with a 'single golden molecular bullet'. The first flaw is appeared relatively recently when problems of resistance to antimicrobial and anticancer drugs became apparent. The multifactorial nature of many complex diseases, such as diabetes, heart disease, cancer and psychiatric disorders, is also important. Most of these diseases cannot be ascribed to single genetic or environmental changes but arise from a combination of genetic, environmental or behavioral factors.[28]

East always believed that complex diseases are best treated with complex combinations of botanical and non-botanical sources that further adjusted to the individual patient and the stage of the disease. This approach is best articulated and developed in traditional Chinese and Ayurvedic medicinal systems. Ostensibly, plants have adapted a similar strategy in their biochemical warfare with pathogens, which are the main causes of plant disease and death. Relying on a single antibiotic to stop pathogens would probably be evolutionarily suicidal for plants because a resistance would develop. Plants also produce diverse groups of antimicrobial compound to prevent the development of resistance have been documented. For example, *Berberis fremontii* Torrey produces both antimicrobial berberine alkaloids and inhibitors of a bacterial multidrug-resistant pump that strongly potentiate the activity of berberines.[29] Interaction between different molecular components can also be required for an optimal therapeutic effect of plant extracts. The root extract of a

Tripterygium wilfordii Hook F has been used historically as a traditional Chinese medicine to treat rheumatoid arthritis, an observation recently supported by a Phase I/II double-blind, placebo-controlled trial[30] in the USA.[31] Although the main active ingredient of this extract, triptolide, has been identified, it was shown to be toxic unless given as a part of the root extract, suggesting that other unidentified extract components increase its safety and, possibly, efficacy.[32]

The future of botanical drugs depends on two factors:

1. Sustaining a favorable regulatory environment and developing technologies for the efficient discovery,
2. Development and manufacture of botanical drugs.

Majority of botanical drugs under development are derived from ethno botanical sources. Through NDA processing common botanical dietary supplements are also upgraded to give ethical drug status.

New creative and innovative technologies are needed for new botanical drug discovery. Production of botanical drugs will require genetically uniformed monocultures of source plants grown in fully standardized conditions to assure biochemical consistency and to optimize safety and efficacy in every crop. Fully controlled greenhouse-based cultivation systems developed for high quality year-round vegetable production are much more suitable for the future production of botanical drugs.

Botanical dietary supplements–also called botanical nutraceuticals or herbals are defined as plant-derived materials with medical benefits. Botanical supplements can be marketed with few regulatory impediments, providing that disease prevention; curing or detection claims are not made. Instead, so-called structure–function claims on products can relate to enhancing or maintaining normal physiological functions of human body. Cosmeceuticals represent a relatively small segment of the total botanical supplement market.

Although the general public often considers botanical supplements natural and safe alternatives to conventional synthetic pharmaceuticals, there is relatively little scientific evidence behind this belief. In 1999 the global market for herbal supplements exceeded US$2.7 in the rest of Asia and US$3 billion in North America.[33] The demand for dietary supplements is driven by a variety of factors that include an aging population with substantial disposable income, a growing trend to self-medicate, mistrust in the conventional medical establishment, and the perception that natural is healthy and that plant products are safe. These trends suggest the growing demand for supplements and functional foods, provided that quality standards and efficacious new products are introduced. The growth of dietary supplements will be hastened by the dramatic advances in human genetics that have culminated in the recent sequencing of the human genome,[34] which will soon allow the reliable assessment of the risk of life-threatening diseases long before their onset. The medical community is not fully prepared for this challenge because it uses monitoring as the main approach to high-risk groups.

Botanical supplements and functional foods might hold the promise for a proactive preventive strategy that follows genetic identification of a medical risk. Botanical supplements are frequently and rightly criticized for poorly proven efficacy and safety, lack of standardization and quality[35] standards,[36] and potential interactions with prescribed drugs.[37] Nevertheless, data from clinical trials performed with the top selling botanical supplements suggest that most are at least mildly effective against a specific[38] indication.[39] Unlike ethical drugs, the active ingredients of very few botanical supplements have been fully characterized. This shows that therapeutic effects of many botanical supplements are the result of the combined effects of many compounds.

Table 2

System	*Protein*	*Expression*
Whole Plant (Cytosolic)	HbsAg, vaccine collagen [65]	.007% TSP 01 mg/g DW
Cellular Component		
Vacoule	scFv, Hapatitis B [67]	0.032% TSP
	sIgA/G [68]	not reported
ER	scFv, cutinase [69]	1% TSP
	scFv, T84.66 [70]	29μg/g FW
	scFv, ABA [71]	6.8% TSP
Tissue-specificity		
Seed	avidin [74]	6% TSP
	hirudin [61]	1% FW
Tuber	scFv, oxalozone [75]	2% TSP
Root	IgM, RKN [76]	0.003% TSP
Fruit	RSV-F protein [77]	not reported
Chloroplast	Somatotropin [78]	7% TSP
Viral	a-trichosanthin [79]	2% TSP

Pharmaceutical are now engaged in recombinant proteins technology, such as production of antibodies, vaccines, regulatory proteins and enzymes. With dozens of proteins in clinical development today there is a substantial shortage of industrial capacity to manufacture future recombinant drugs.[40] During the past decade, plants have emerged as promising biopharming systems for commercial production of pharmaceutical proteins (Table 2). Tobacco was the first plant to express a recombinant antibody in 1988,[41] with further confirmation in 1989.[42] Advantages offered by plants include low cost of cultivation and high biomass production, relatively fast 'gene to protein' time, low capital and operating costs, excellent scalability, eucaryotic post-translational modifications (i.e. glycosylation, folding and multimeric assembly), low risk of human pathogens and endotoxins and a relatively high protein yield. These advantages are potentiated by the ease of plant transformation through particle bombardment, electroporation, *Agrobacterium*-mediated transformation, or infection with modified viral vectors.[43]

Plants are generally considered to be low-cost, safe and relatively fast alternatives to many existing manufacturing systems. Most major groups of human pharmaceutical proteins have been produced successfully variety of crops and model systems (e.g. maize, rice, wheat, soybean, tomato, potato, mustard, oilseed rape, turnip, alfalfa, banana, tobacco and *Arabidopsis*) using stable nuclear and plastid transformations, as well as transient expression systems[44] such as viruses.[45]

Plants can successfully perform the post-translational modifications required by the majority of pharmaceutical proteins under development. Heterologous proteins requiring these modifications are usually retained in the endoplasmic reticulum (ER) using the C-terminal ER retention sequence KDEL or targeted into the protein secretion-modification pathway that delivers the recombinant proteins into the intercellular spaces (apoplast) via the ER and Golgi apparatus. Both strategies significantly enhance protein[46] expression.[47] The highest yield of recombinant protein in plants is achieved by chloroplast expression or possibly by transient viral expression. However, plastid expression suffers from the lack of glycosylation and corrects folding of multimeric proteins, whereas viral expression is problematic for complex proteins and could introduce safety hazards. Seeds (corn) and tubers (potatoes) provide excellent long-term storage compartments for recombinant proteins before purification, but have lower protein yield.

Downstream protein purification is often as expensive as the biomanufacturing and should never be overlooked in the total 'cost of goods' equation. At least two approaches have been used successfully to lower the cost of downstream purification of plant-produced proteins: oleosin-fusion technology for heterologous proteins produced in oilseeds,[48] and rhizo- and phyllo-secretion platforms based on continuous, non-destructive recovery of a target protein from plant[49] exudates.[50] The latter also offers the advantage of continuous protein production that integrates the biosynthetic potential of a plant over its lifetime and might lead to higher protein yields than single harvest and extraction methods. With no plant-produced protein drug on the market today, and only a few in clinical trials, one can only hypothesize that plants will provide substantial benefits and savings for protein manufacturing over the alternative systems. Nevertheless, the arguments favoring plants are appealing and a growing number of companies are trying to commercialize recombinant protein manufacturing in plants, with most concentrating on pharmacological applications. It is likely that no single ideal system will ever emerge for the manufacturing of every recombinant protein, each system having distinct advantages and disadvantages.[51] Emphasis has shifted from the vector construct to the host organism. Biotechnology.

Conclusions

Plants are arguably poised for a comeback as sources of human health products. The hopes for this comeback are rooted in the unique and newly appreciated properties of phytochemicals *vis-à-vis* conventional NCE-based pharmaceuticals and are based on the:

- Enormous propensity of plants to synthesize mixtures of structurally diverse bioactive compounds with multiple and mutually potentiating therapeutic effects;
- Low-cost and highly scalable protein and secondary metabolite biomanufacturing capacity of plants;
- Diminishing return of the single NCE approach to drug discovery and disease treatment and prevention;
- Cost limitation on the chemical synthesis of complex bioactive molecules; and
- Perception that because of the history of human use and co-evolution of plants and humans, phytochemicals provide a safer and more holistic approach to disease treatment and prevention.

Although the above properties have been known for a long time, the ability to better exploit the uniqueness of plant therapeutics was acquired only recently as a result of the dramatic advances in metabolic engineering, biochemical genomics, chemical separation, molecular characterization and pharmaceutical screening. A challenge for phytochemical-based botanical therapeutics is to integrate the ability to identify and genetically manipulate complex biosynthetic pathways in plants with better characterization of genetic targets for the prevention and treatment of complex diseases. Similarly, an important challenge is the development of discovery, validation and manufacturing technologies that are compatible with multifunctional phytochemical mixtures. The discovery, development and manufacturing of botanical therapeutics, either isolated from plants or delivered as food constituents, is likely to be a major area of plant biotechnology expansion in the twenty-first century. It is likely that botanical drugs, dietary supplements and recombinant protein production will have a major role in this expansion. However, plants are unlikely to recapture their previously dominant role as sources for NCE discovery. It is harder to make predictions for the area of functional/medicinal foods. Consumer and regulatory acceptance of drugs delivered in foods will be a decisive factor in determining the future of these products. Whereas the increasingly holistic approach to treatment and prevention will continue to emphasize food, consumer food choices might still be determined by individual taste predilection, habits and cultural heritage rather than by health benefits. Thus, pills taken with 'not-so-healthy' meals might continue to be a preferred vehicle for delivering disease treatment and prevention. The growth of botanical therapeutics might add more value to world agriculture than the more conventional application of plant biotechnology for yield enhancement. Plants are by far the most abundant and cost-effective renewable resource uniquely adapted to complex biochemical synthesis. The increasing cost of energy and chemical raw materials, combined with the environmental concerns associated with conventional pharmaceutical manufacturing, will make plants even more compatible in the future. Crops that will benefit the most include tobacco and corn as the major recombinant protein manufacturing crops and many minor crops and medicinal plants that will become sources of future botanical therapeutics. Farmers that adapt to growing crops for health, rather than calories, will profit from greater margins and higher values enjoyed by the health industry and, as a result, the planet could become greener and more healthy.

Notes and References

1. Swerdlow, J. (2000), Nature's Medicine. Plants That Heal, National Geographic Society.
2. Pierpoint, W.S. (1994), Salicylic acid and its derivatives in plants: Medicines, metabolites and messenger molecules. Adv. Bot. Res. 20, 163–235.
3. Business Communications Company, Inc. (1998) Plant-Derived Drugs: Products, Technologies and Applications. Study RB-121.
4. Farnsworth, N.R. (1988), Screening plants for new medicines. In Biodiversity (Wilson, E.O., ed.), pp. 83–97, National Academy Press.
5. Cordell, G.A. (2000) Biodiversity and.drug discovery. Asymbiotic relationship. Phytochemistry 55, 463–480.
6. Newman, D.J. *et al.* (2000) The influence of natural products upon drug discovery. Nat. Prod. Rep. 17, 215–234.
7. Farnsworth, N.R. and Morris, R.W. (1976) Higher plants–the sleeping giant of drug development. Am. J. Pharm. Educ. 148, 46–52.
8. Farnsworth, N.R. (1988), Screening plants for new medicines. In Biodiversity (Wilson, E.O., ed.), pp. 83–97, National Academy Press.
9. Farnsworth, N.R. (1988), Screening plants for new medicines. In Biodiversity (Wilson, E.O., ed.), pp. 83–97, National Academy Press.
10. Rates, S.M.K. (2001), Plants as sources of drugs. Toxicon 39, 603–613.
11. Adang, A.E. and Hermkens, P.H. (2001) The contribution of combinatorial approaches to lead generation: An interim analysis. Curr. Med. Chem. 8, 985–998.
12. Schreiber, S.L. (2000), Target-oriented and diversity-oriented organic synthesis in drugdiscovery. Science 287, 1964–1969.
13. Clark, D.E. and Pickett, S.D. (2000), Computational methods for the prediction of drug-likeness. Drug Discov. Today 5, 49–58.

14. Gentry, A.H. (1993), Tropical forest biodiversity and the potential for new medicinal plants.ACS Symposium Series Washington, DC American Chemical Society 534, 13–20.
15. Mendelson, R. and Balick, M.J. (1995), The value of undiscovered pharmaceuticals in tropical forests. Econ. Bot. 49, 223–228.
16. Shu, Y.Z. (1998), Recent natural products based drug development: A pharmaceutical industry perspective. J. Nat. Prod. 61, 1053–1071.
17. Bolten, B.M. and De Gregorio, T. (2002), Trends in development cycle. Nat. Rev. 1, 335–336.
18. *Ibid.*
19. Cordell, G.A. *op cit.*, p. 463–480.
20. Ebel, J. and Cosio, E.G. (1994), Elicitors of plant defense responses. Int. Rev. Cytol. 148, 1–36.
21. Facchini, P.J. (2001), Alkaloid biosynthesis in plants: biochemistry, cell biology, molecular regulation, and metabolic engineering applications. Annu. Rev. Plant Physiol. Plant Mol Biol. 52, 29–66.
22. Dixon, R.A. and Paiva, N.L. (1995), Stress-induced phenylpropanoid metabolism. Plant Cell 7, 1085–1097.
23. Trapp, S. and Croteau, R. (2001), Defensive resin biosynthesis in conifers. Annu. Rev. Plant Physiol. Plant Mol. Biol 52, 689–724.
24. Turlings, T.C.J. and Tumlinson, J.H. (1992), Systemic release of chemical signals by herbivore-injured corn. Proc. Natl. Acad. Sci. U. S. A. 89, 8399–8402.
25. Darvill, A.G. and Albersheim, P. (1984), Phytoalexins and their elicitors–a defense against microbial infection in plants. Annu. Rev. Plant Physiol. Plant Mol Biol. 35, 243–275.
26. Dixon, R.A. (1986), The phytoalexin response: Elicitation, signaling, and control of host gene expression. Biol. Rev. 61, 239–291.
27. U.S. Department of Health and Human Services Food and Drug Administration Center for Drug Evaluation and Research (2000) Guidance for Industry Botanical Drug Products. Online: http://www.fda.gov/cder/guidance/1221dft.htm#P131_3293.
28. Kibertis, P. and Roberts, L. (2002), It's not just thegenes. Science 296, 685.
29. Hsiang, P. and Lewis, K. (2000), Synergy in medicinal plants: anitimicrobials potentiated by multidrug pump inhibitors. Proc. Natl. Acad. Sci. U. S. A. 97, 1433–1437.
30. Tao, X. *et al.* (2001), A Phase I study of ethyl acetate extract of the Chinese antirheumatic herb Tripterygium wilfordii Hook F in rheumatoid arthritis. J. Rheumatol. 28, 2160–2167.
31. Tao, X. *et al.* (2002), Benefit of an extract of Tripterygium wilfordii Hook F in patients with rheumatoid arthritis: A double-blind, placebo-controlled study. Arthritis Rheum. 46, 1735–1743.
32. Su, D. *et al.* (1990), Comparative clinical study of rheumatoid arthritis treated by triptolide and an ethyl acetate extract of Tripterygium wilfordii. Chin. J. Mod. Dev. Trad. Med. 10, 144–146.

33. Glaser, V. (1999), Billion-dollar market blossoms as botanicals take root. Nat. Biotechnol. 17, 17–18.
34. Venter, J.C. *et al.* (2001), The sequence of human genome. Science 291, 1304–1351.
35. Osowski, S. *et al.* (2000), Pharmaceutical comparability of different therapeutic Echinacea preparations. Res. Compl. Nat. Classical Med. 7, 294–300.
36. Kressmann, S. *et al.* (2002), Pharmaceutical quality of different Ginkgo biloba brands. J. Pharm. Pharmacol. 54, 661–669.
37. Izzo, A. A. and Ernst, E. (2001), Interactions between herbal medicines and prescribed drugs: A systematic review. Drugs 61, 2163–2175.
38. Ernst, E. (2002), The risk-benefit profile of commonly used herbal therapies: Ginkgo, St. John's Wort, Ginseng, Echinacea, Saw Palmetto, and Kava. Ann. Intern. Med. 136, 42–53.
39. Kroll, D.J. (2001), Concerns and needs for research in herbal supplement pharmacotherapy and safety. J. Herbal Pharmacother. 1, 3–23.
40. Garber, K. (2001), Biotech industry faces new bottleneck. Nat. Biotechnol. 19, 184–185.
41. During, K. (1988), Wundinduzierbare Expression und Sekretion von T4 Lysozym und monoklonalen Antikorpern in Nicotiana tabacum. Doctoral Dissertation, University of Koln, Germany.
42. Hiatt, A. *et al.* (1989), Production of antibodies in transgenic plants. Nature 342, 76–78.
43. Fischer, R. and Emans, N. (2000), Molecular pharming of pharmaceutical proteins. Transgenic Res. 9, 279–299.
44. Giddings, G. *et al.* (2000), Transgenic plants as factories for biopharmaceuticals. Nat. Biotechnol. 18, 1151–1155.
45. Daniell, H. (2001), Medical molecular pharming: production of antibodies, biopharmaceuticals and edible vaccines in plants. Trends Plant Biol. 6, 219–226.
46. Conrad, U. and Fiedler, U. (1998), Compartment-specific accumulation of recombinant immunoglobulins in plant cells: An essential tool for antibody production and immunomodulation of physiological functions and pathogen activity. Plant Mol. Biol. 38, 101–109.
47. Firek, S. *et al.* (1993), Secretion of a functional single-chain Fv protein in transgenic tobacco plants and cell suspension cultures. Plant Mol. Biol. 23, 861–870.
48. Parmenter, D.L. *et al.* (1995), Production of biologically active hirudin in plant seeds using oleosin partitioning. Plant Mol. Biol. 29, 1167–1180.
49. Borisjuk, N.V. *et al.* (1999), Production of recombinant proteins in plant root exudates. Nat. Biotechnol. 17, 466–469.
50. Komarnytsky, S. *et al.* (2000), Production of recombinant proteins in tobacco guttation fluid. Plant Physiol. 124, 927–933.
51. Hodgson, J. (1993), Expression systems: A user's guide. Emphasis has shifted from the vector construct to the host organism. Biotechnology.

TESTING FAITH, WRESTLING WITH MYSTERY

Vineeta Tiwari

'FAITH', the dictionary meaning of the word is 'confident belief in the truth'. It includes within it some or all of the following beliefs–belief in God, belief in the immortality of the soul, belief in an after life, belief in heaven and hell and so on. It also can be termed as 'religious belief'.

Have you ever been in a situation when your faith is tested? Have you ever wondered where God's is when you are in difficulties even though you have been crying out for him to rescue and have seen none? I have, and you probably have too. That is because the nature of faith is empirical. It can not be imposed by influence but can be felt by our own experience. It is very personal. It concerns not 'God' but 'my God'. "I don't have a religion. I believe in a God. I don't know what it looks like but it's My God. My own interpretation of the supernatural"–Jennifer Aniston[1].

It is true that during all these trials when your faith was being tested somehow you were expecting from God. This means in all these situations your faith depends upon the fulfillment of your own desire. That means 'hope' is one condition for faith. Now think just reverse of it i.e. what will happen if God allows you to go through these difficulties to test your faith? Will it withdraw you from your spiritual walk, or will you continue on in faith in spite of your circumstances and do exactly what the Bible says: "The just shall live by faith?"[2] To live by faith means that you trust the God in all circumstances.

In my view what inspires people to strive towards spiritual ideal is not reason but an emotional conviction. Reason functions as means of testing and clarifying our convictions and faith achieves its decisive confirmation through experience. Faith has long been a puzzle for science. By definition faith demands belief without a need for supporting evidence, a concept that is not more opposed to the principles of scientific inquiry. In the eyes of the scientists, "an absence of evidence reduces belief to a hunch."[3] Before science could make its magnificent achievements, religion and theology had much captured the human mind. But as science has gone on advancement, religion has suffered a gradual setback.

As conviction of faith is empirical in nature, I gathered some statements of great people and thinkers to know what they experienced and said about faith. Swami Vivekananda, a great thinker of modern India quotes faith as–'sincere belief in a concept, and consequent eagerness to reach it'. He said, "Experience of others is faith for the beginner. A father tells his son - that the mango is sweet–and the child eats it and confirms its sweetness. Here 'faith' (in father about sweetness of mango) got converted into 'knowledge' relatively quickly. Another example, Copernicus came to the conclusion that the earth moves the sun; not the other way round. A few people believed in him in the beginning. However, the truth of his proposition was confirmed later, and thus became knowledge for everyone. Here faith took longer time to get realized."[4]

Faith in view of Mahatma Gandhi–"If we exist, if our parents and their parents have existed then it is proper to believe in the parent of the whole creation ... An atheist might floor me in a debate. But my faith runs so very much faster than my reason that I can challenge the whole world and say, 'God is, was and ever shall be' ... There is an indefinable mysterious power that pervades everything. I feel it, though I don't see it."[5]

Rabindranath Tagore, a great poet and philosopher says–"With the advancement in science and technology human life has been endowed with much comfort. Achievement of science and technology have unfolded new horizon for humanity. But we should not forget that, 'science has liberated man from much of the tyranny of the environment but has not freed him from the tyranny of his own nature. So the need of religion is beyond argument'."[6]

Hick in his writings says–"If a man accepts a given belief that is widely accepted he need not to explain the fact because it is widely accepted. But if he does not accept it, he may well come to wonder why so many people do accept it. From ancient times there have been many attempts on the part of religious skeptics to answer this sort of question. The simpler answer is–These beliefs are a survival from the earliest human attempts to explain natural phenomena."[7]

There are many theories that prove the existence of God. On the other side our atheistic view compels us to disbelieve all those facts and theories. An optimistic view encompasses not only trust but also awe, dread, dismay, resentment and even hatred. This is my own experience. On the other side atheistic or pessimistic view consequently follows downfall of our own morals. If you are an atheist you can never think positively and ultimately it will ruin you. Without faith we would never take up a challenge. "Better go without belief forever than believe a lie! Merely shows his own preponderant private horror of becoming a dupe".[8] Reason could not prove God, but still there must be a God to meet the demands of morality. Morality requires him as its very presupposition.

None of the proofs is able on valid logical grounds to prove the existence of such a being.

Max Planck (a physicist) says, "Science can not solve the ultimate mystery of nature ... scientific statements are valid only within the framework of strictly limited relations. While they study the atoms and electrons, they do not and cannot say anything about the creation of atoms and electrons."[9] Even Darwin's theory of evolution is based on this faith. We may also refuse to accept his 'accidental variations' theory that says changes are due to the environmental conditions. The same parents living in the same conditions and environment may produce a genius at one birth and an idiot at the next.

Though 'knowing the truth' has been a universal topic of research yet nobody could reach the fact. Swami Vivekananda says –'test everything in the crucible of your reasoning then accept it'. He himself has tested and analysed every religion and praised the God in his writings. I too have come across doctors and scientists who claimed to have experience and felt the power of a Supreme Being.

Conclusion

After analysing all these facts that may suffice to approve the existence of God, what are we to doubt? Encountering that doubt is only like a personal encounter and proves that 'testing faith is really like wrestling with the mystery, the mystery unsolved yet and will ever be though I recommend to test your faith because it empowers you in someway. And as bestowed with the power of brain, our curiosity will lead us some day, some way to the truth definitely. This is my faith.

Notes

1. A quotation from TOI Feb 2006.
2. Internet sources
3. Internet sources
4. Internet sources
5. M. K. Gandhi. My Religion. Navajivan Publishing House (Ahmedabad, I Edn., 1955).
6. Chandra Mohan Das. The Philosophy of Rabindranath Tagore. Deep & Deep Publications. (New Delhi, 1996).
7. John Hick. Faith and the Philosophers. St. Martin's Press. (New York, 1964).
8. William James. The will to believe human immortality and other essays in popular philosophy. Dover Publications Inc. (1956).
9. Chandra Mohan Das. The Philosophy of Rabindranath Tagore. Deep & Deep publications. (New Delhi, 1996).

BIBLICAL PERSPECTIVE TOWARDS ALARMING ECOLOGICAL SCENARIO

Samuel Richmond Saxena

Introduction

Global warming! Tsunami waves! Earthquakes, Floods! Droughts! Famine! ...the list goes on and on. Is there an order to control these destructive elements from the surface of the earth? What is the destiny of humankind? God created the universe so in what ways does He express concern for His creation? In the present post-modern technical world of academicians, economist, politicians, sociologists, theologians and other religious leaders, ecology has become a burning and fervent issue of discussions which has added to the vast body of eco-literature. Instead of sympathy, love and care towards nature, we humans are ruthlessly destroying it. This is a wake up call for every human being who has not only love for God but also for His creation. God created man in His own image and appointed him to look after the creation as a steward. Are the present Christians abiding the Biblical laws regarding the care of God's creation? Some people may think, "Christians? Why should they care about the environment?" And unfortunately, many of us may stream our thinking in the same line. "What does my faith have anything to do with a concern for nature?"[1] Christians are not expected to do anything about the environment, and thus they do not. In this present eco-crisis the Bible offers a solution which is relevant in the present context. Therefore, there is an urgent need to address such a social issue with a Biblical perspective.

Nature and its Values

Since creation, human beings have been a part of nature and cannot be considered in isolation. The beginning of the civilization, in fact, coincided with the use of fire and other tools to modify environment. Therefore, from pre-historic times to the present, the civilization has been recognizing nature for worship, myths, rituals, and religious scriptures. *Prakrati* (Nature) a fundamental concept in Indian philosophy is a goddess of a personified energy or wife of a deity and is generally considered as identical with the Supreme Being.[2] The continuous dependence of human on nature for providence has revealed the values attached to it. No one can deny or ignore the magnificence and majesty of nature which elicits a sense of awe in us. And God has given us the wisdom to recognize these values (of nature) by appreciating its beauty, getting providence out of it and caring and sustaining it for healthy environment. We because of our anthropocentric character are more inclined towards the conventional values[3] rather than adopting green values (contrary to conventional value) of nature. Apart from the above two values there are many more which are attached to the nature i.e. economical value, health value, re-creative value and aesthetic value. The eco-crisis movement has once again brought the values of nature in light. So how do we value nature in the present century when there is a great threat to it?

Ecological Crisis

The proposed construction of world's tallest Buddha statue on a rock face near Bangalore has angered several environmentalists, rock climbers, and number of local people. They have called this act "a gross violation of environmental norms." Environmentalists claim that this construction will bring out shrinkage inhabitant of rare birds, endangered sloth bear and cattle grazing. After calculation, an estimated 2000 acres land used for this project will force many to leave their ancestral land.[4] Sometimes it appears strange that how government can allow such projects to start and overlooks its consequences to severe threat to common people (especially the poor) and the species of flora and fauna.

The increase in the global warming, rapid destruction of ozone layer which has reached alarming proportion, with 30% depletion by 1984.[5] The extinction of three species of plants and animals each day, poaching racket of tigers and leopard hides,[6] the entering of French warship with toxic wastes in Indian waters,[7] decrease of the African elephants by 70% within the span of ten years, killing of 10 millions of reptiles for their skins every year, the removal of 1,00,000 square kilometers of primary forest each year,[8] death of 4,000 birds on the shore of northwest Estonia due to oil pollution in the last few weeks are enough to alarm the nonstop ill-treatment and unethical behavior of humans towards the nature.[9] Due to the global warming, as the planet heats up, more and more areas across the world receive more rain than ever before, while paradoxically, many areas receive less or no rainfall at all.[10] The present data reveals that the rate of Carbon Dioxide in the atmosphere has increased in comparison to last many years.[11] The prediction of Scientists, that average global temperature will rise by 1.5 to 4.5°C by year 2030 will cause the ice cap in the Antarctica and Artic to melt down and heated surface waters of the oceans will rise by 6 feet further up which will drown many costal areas and islands.[12] Due to the global capitalization the multinational companies are heading towards the profit making which is resulting into ecological disaster.[13] Ultimately, the whole ecosystem is in a big threat and has become an alarming issue among the ecologists as well as theologians.

When humans try to circumvent the laws of nature, they usually destroy the natural environment that sustains them and thus due to technological achievement, human seems to depend less on the natural environment for their daily growth. Rev Richard Howell clearly writes, "Nature has an economical value and the environmental crisis needs to be treated as an economic problem. Nature also has health value and the protection of the environment needs to be placed within the framework of preventive health care."[14]

Therefore if the natural values are recognized and the providence understood then we can start an effort to control this rising ecological crisis and provide a sustainable ecosystem for the next generation. How long will we fool around with God's creation like this? Without realizing the providence and the purpose of this magnificent creation which God has created for us? What does the religion have to do with space travel, genetic engineering or a depleted ozone layer? Are Christians so concerned with preserving biblical truth that we have little to offer those who are the victims of ecological crisis?

God and Creation

Every religion of this world has different perspective about the relationship of God with His creation. But the uniqueness of relation between God and his Creation can be seen in the Bible. Some of us must be wondering why am I bothering about creation of God? God created this entire universe in six days and on the sixth day when he created everything, then He placed human beings as the crown of creation. It took six days for God to create excellent and a balanced environment for them. All the gases in the air, chemical elements & compounds of the earth and water were balanced. Everything was fresh, clean and spanking new without deficiency. Imagine if God would have created human beings on the very first day? What would have happened? But God felt that human beings needs surrounding and proper environment to survive, and to multiply. He admired his handiwork by saying "good" and the completeness of the creation came to the halt when God created Human beings in His own image "And God saw....that it was really very good." (Gen 1:31). Therefore God never created anything evil but "good." When every thing was finished He blessed the entire creation. Mostly, people think that God has nothing to do with His creation. He is superior and has confined himself to some part of galaxy beyond human reach. Biblical doctrine of creation asserts that sovereign God who is the *alfa* and *omega* like an impersonal God of all existence does not exist some where in the galaxy.[15] Bible asserts that God created the universe

Creatio ex nihilio out of nothing, that shows nothing existed before creation except God and rejects the doctrine of dualism.[16] The opening verse (Gen. 1:1) immediately confronts the reader with the eternal and superior God who is above all (*transcendent*), but still no one can ignore His involvement in creation (*immanent*). In Christian tradition there's a term *Creatio Continua* which describes that God holds the creation and continuously interacts with it. Sometimes the extreme formulation of the involvement of God with his creation has been misunderstood and ended into the pantheism (doctrine which reduces the reality into total identity) by many religions.[17] He is not the part of creation but he has made it and rules over it.[18] The sovereign Lord speaks and creation reacts, this action clearly demarcates between the Creator and his creation. The form and filling solution reveals the artistic work of God and projects the Biblical creation as unique and perfect. The first three days show God's action in forming and the last three days to fill these forms.

Acts of Forming	*Acts of Filling*
Day 1: Light/Darkness	Day 4: Lights of Day & Night
Day 2: Sea/Sky	Day 5: Creatures of Sea and Sky
Day 3: Fertile Earth	Day 6: Creatures of Earth[19]

His admiration towards His creation reveals the labor and concern for His creation. And placing human beings as stewards to take care of this entire planet shows His involvement. Job 12:10 says that even, human beings, animals and plants depend on God. But human beings after the fall are continuously distorting the outstanding artistic work of God. Psalm 8 is a *hymn of praise* and it may be classified as *creation hymn*. David in this hymn adores the work of the creator and reminds the responsibility of Jewish community as a good steward to His creation. Only humankind can alone reflect divine qualities and have communion with God. Neither animate nor inanimate has this privilege of sharing. At last, this passage gives a close connection between God as King and God as Creator.[20]

The fall of Creation

Since the fall of creation there's a rift between Creator and creation which resulted in sin and death (Rom. 5:12). Through disobedience and failure of the first human couple the whole ecosystem became imbalanced and the entire planet was cursed by God (Gen 3:17), because of this curse the materialistic world and the flesh is destined to pass away and ultimately be destroyed. By now it should be understood that spirit alone has value till eternity. The popular gospel song sung by Jim Reeves conveys, "This world is not my home, I'm just a passing through. My treasures are laid up somewhere beyond the blue. The angels beckon me from heaven's open door. And I can't feel at home in this world anymore."[21]

Many Christians feel that why to take care of this creation when it would collapse? Interestingly many of us fail to understand that even after the fall, human dominion still continues over God's creation under proper obedience. According to Odum both the terms ecology and economics consist of Greek word *oikos* meaning "household." Former is the "study of the household" and latter is "the management of the household." Therefore ecology is the study of environmental house which consists of all organisms and their relation with each other while economics (under ecological terminology) is the management of the entire ecosystem through the human ruler-ship.[22] Surprisingly, the two Hebrew words in Gen. 1: 28 i.e. *kabas* (subdue) and *radah* (dominion) have been misinterpreted by many Old Testament Bible scholars as to "trample" or to "exploit" and due to this understanding the management has turned into exploitation.[23]

While Gordan Wehnem writes "Because man is created in God's image, he is king over nature. He rules the world on God's behalf. This is of course no license for the unbridled exploitation and subjugation of nature. Ancient oriental kings were expected to be devoted to the welfare of their subjects, especially the poorest and weakest members of society. By upholding divine principles of law and justice, rulers promoted peace and prosperity for all their subjects. Similarly, mankind is here commissioned to rule nature as a benevolent king, acting as God's representative over them and therefore treating them in the same way as God who created them."[24]

We biblical stewards are responsible to manage and to maintain harmony among the organisms with obedience. But we have failed to shoulder the task of caring and therefore a severe mismanagement has continued since ages till today. How long will it continue? Our good God values human lives and His embodiment in the form of human flesh as Jesus depicts His nature. The incarnation and the resurrection of Jesus Christ proves that God has not forsaken nature. Through incarnation in Jesus, God reaffirmed his concern and love for nature. His resurrection affirms of entering into new ecosystem i.e. new heaven and new earth. Scripture speaks about the sanctification of the whole creation through Jesus Christ.[25]

Christ as the Liberator

If we stretch back to the history of Church we'll find that there were many committed Christians who had great concern for the environment. St. Francis of Assisi, the 12th century monk who founded the Greyfriars, was well known for his love of nature and animals. Another Francis, modern-day theologian Francis Schaeffer, has also written much about Christianity and the environment. At last no one can deny the contribution of great liberation theologians such as Gustavo Gutierrez[26] and Leonardo Boff from Latin America on ecological issues. Along with them number of Latin American liberation theologians have developed a branch of study called eco-theology.[27]

After the II World War disaster and the dropping of atom bombs on Nagasaki and Hiroshima, social lives were ruined and the victims were confused and many questions were left unanswered. It was the first time that even theology was about to cease and people gave up hope. Many atheists were challenging God of Christian theism and accused them that their God has drifted away from human problems and is only seen in supernatural disasters. In that era German theologian–Jurgen Moultmann introduced the *Theology of Hope* where he showed Christ as the liberator in the midst of all the disaster and puts Christian faith with eschatology as the central theme.[28] He claimed that Christ is the "ground of salvation and existence" of the cosmos, through the divine power of the Holy Spirit. He argued that Christ is the only way who can liberate humanity from its misery and conserve nature from total destruction. "Resurrection has become the universal 'law' of creation, not merely for human beings, but for animals, plants, stones and all cosmic life system as well. The raised body of Christ therefore acts as an embodied promise for the whole creation."[29] Hence every organism in creation is precious in God's sight, and Christ's death on the cross is for reconciliation of the world. Therefore *reconciliation* of the cosmos is the re-establishment of his *righteousness* and *justice* to the cosmos.

Future Glory

Paul indicates in Romans 8:18-25 that due to human disobedience the nature has suffered and its suffering is not by its choice. Whole creation was subjected to futility due to the fall (8: 20) and longing for deliverance. 'Creation is groaning' clearly be felt in symptoms such as inexplicable sufferings of human beings and other flora and fauna, floods, cyclones, droughts and many other as mentioned earlier. One day this longing will be realized (cf. Revelation 22: 3). God has judged it by his righteousness and will redeem it by his grace.[30] Hence the hope of the material creation is resting upon the future glory which will be finally revealed (restored) in His time. All the human greed, exploitation, destruction, crisis will come to an end. 'He will wipe every tear from their eyes. There will be no more death or mourning or crying or pain, for the older order of things has passed away.' (Rev 21: 4). A Christian believer, without wasting a moment, must immediately come into action as responsible members through the process of restoration and reconciliation of nature and humanity with keen anticipation and a part of the next great event i.e. the fulfillment of vision of the New Heaven and the New Earth through the second coming of our Lord and Savior Jesus Christ.

Now the question is for the Christian believers how they should work for their creation? Col. 3: 23 says, 'Whatever you do, work at it with all your heart, as working for the Lord not for men.' Most of our actives are pleasing our greed which is seriously detrimental to the nature. As good.

stewards, we have to conscientize people towards this eco-crisis and make them responsible to this God's creation. Long back an American scientist had mentioned in one of the science journal 'that due to the rapacious attitude of human towards nature, the nature has started responding negatively towards human and the natural calamities recorded in the last two centuries are more than the calamities recorded in the entire history of the planet earth.' Therefore, we have to work hard for the betterment of the planet with responsibility and accountability to God because whatever work we do, we do ultimately for the glory of God.

Notes

1. Allen Yeh, Seminar Presentation on *Ecotheology and the Self-Colonization of Latin America* (D.Phil. candidate-Department of Theology, Oxford University), p. 1.
2. Joseph Peruma, The Motherly Earth: Ecological Ethics of Human-Nature Relationship (Bangalore: Claretian, 2002) p. 91.
3. Conventional Values: Humans are separate from nature, nature can and should be exploited and dominated for human benefit, and we can and should use the laws of nature (scientific laws) to exploit and use it.
4. Stephen David, 'A Rock and a Hard Place' *India Today*, June 27th 2005. Vol. 30/25. p.78-79.
5. D.C. Srivastava (ed.), Readings in Environmental Ethics: Multidisciplinary Perspective (Jaipur: Rawat, 2005), p. 91.
6. B.K. Sharma, 'The Killing Game' *Sanctuary*, August 2005, Vol. 25/4. P. 78-80.
7. Web feature in India Today, January 30th 2006, Vol. 31/4 p. 9.
8. D.C. Srivastava (ed.), *Readings in Environmental Ethics: Multidisciplinary Perspective* (Jaipur: Rawat, 2005), p. 91.
9. 'Oil Pollution,' *THE HINDU*, Young World (ed.), Feburary 10th, 2006. P. 3.
10. Lakshmy Raman, *'War on Nature: The Case of Gujrat'* in Sanctuary, Vol 25/4 August 2005, p. 66.
11. N.A. Increased Global Pollution Confronts Montreal Meet, *The Hindu*, (Delhi: November 30, 2005), p. 16.
12. K.P. Joy, *Greening of Minds: Environmental Education for Sustainable Future*, (Thiruvalla: Christava Sahitya Samiti, 2001), p. 27.
13. Jacob Parappally (ed), *Theologizing in Context: Statement of the Indian Theological Association* (Bangalore: Dharmaram Publication, 2002), p. 252.
14. D.C. Srivastava (ed.), Readings in Environmental Ethics: Multidisciplinary Perspective (Jaipur: Rawat, 2005), p. 92.
15. Vinoth Ramachandra, *Gods That Fail: Modern Idolatory and Christian Mission*, (London: Paternoster Press, 1996), p. 73.
16. Dualism: God and the material universe eternally existed side by side. There are two forces God and nature and there is an eternal conflict between God and evil of the materiel universe. This theory denies the total lordship or governance of God over creation.

17. Augustine Pamplany, *Theological Mysteries in Scientific Perspective* (Bangalore: Asian Trading Cooperation, 2005), p. 170.
18. Wayne Grudem, Systematic Theology: An Introduction to Biblical Doctrine (Grand Rapids: Michigan, 1994) p. 267.
19. Vinoth Ramchandra, p. 60.
20. O.P. Dwivedi, *World Religions and the Environment* (New Delhi: Gitanjali Pub, 1989). p 77.
21. Ibidi., p. 82.
22. Fundamentals of Ecology, Odum, p. 2.
23. Rajula Annie Watson, *A Development and Justice: A Christian Understanding of Land Ethics* (Delhi: ISPCK, 2004), p. 68-69.
24. Wenham, Gordon J., Word Biblical Commentary, Volume 1: Genesis 1-15, (Dallas, Texas: Word Books, Publisher) 1998.
25. D.C. Srivastava (ed.), *Readings in Environmental Ethics*, p. 98.
26. Father of Liberation Theology.
27. Allen Yeh, Seminar Presentation on *Ecotheology and the Self-Colonization of Latin America* (D.Phil. candidate-Department of Theology, Oxford University), p. 2.
28. Stanley J. Grenz & Roger E. Olson, *20th Century Theology: God & the World in a Transitional Age* (Secunderabad: OM Pub 1992), p. 172-173.
29. Rajula Annie Watson, *A Development and Justice: A Christian Understanding of Land Ethics* (Delhi: ISPCK, 2004), p. 220-223
30. D.C. Srivastava (ed.), *Readings in Environmental Ethics*, p. 98.

RELIGIOUS FAITHS TO PRESERVE ECOLOGY

Peter Amaladoss

Man and his environment are inextricably linked with each other and constantly interacting with each other. This interaction between man and his environment is as old as the appearance of life on this earth; but a disharmony or imbalance in this relationship is a recent occurrence. By a modest estimate, it may be said that the earth came into existence about five billion years ago; life on this earth evolved around three billion years ago; human beings appeared about 0.5million years ago and the modern species of man appeared about 0.15 million years ago. The earliest human civilizations along the river valleys appeared just about 4500 years ago. The spread of human population was extremely slow in the distant past; but it registered significant acceleration in the last three centuries particularly after the Industrial Revolution of the 18th century. The increasing human population and the expanding need to provide it with food, habitation and other essentials of life led to increasing pressure on the environment and also its defiling by a variety of pollutants. It assumed the dimensions of a major crisis by the mid-20th century.

Past and Present

The environment did not pose any problem earlier because man did not interfere with the forces of nature and worshipped nature as his real benefactor and life-sustainer. Whatever environmental problem was there in those days of nature-friendly age was caused by nature itself. Man was a silent spectator; he had neither the inclination nor the tools and will to defy the forces of nature. Thus, the sudden change in the course of a river, the indiscriminate cutting of forest and the subsequent loss of green cover in some areas, flood, draught, fire, cyclone etc. caused environmental crisis even in earlier days. But they were all nature-made and nature alone knew how to overcome those crises. Environmental crisis of our age however is 90% manmade.

Crisis

Before embarking on discussing the crisis as such, it will be worthwhile to analyze what after all is the nature of this crisis. It is just like our breathing process where we constantly inhale and exhale but are not at all aware of the process. Only when some problems arise, we do realize the importance and the necessity of air and the process of breathing we are performing. Like wise, environment is ever present where life exists. Both have perennial co-existence. When this co-existence is balanced then there is harmonious, prosperous, peaceful and serene relationship among the whole creation. This harmonious co-existence or ecological equilibrium ought to be preserved at any cost if life on the planet earth is to survive. Our problem arises due to disturbed and upset equilibrium of the natural environment. The following actions disturb one or more elements of 'natural equilibrium'

- Agriculture, deforestation, land development, creation of manmade lakes, dams and other activities changing the structure of the earth,
- Changing the composition of the biosphere through extraction of minerals, discharge of various substances into atmosphere and into water bodies and on land, and
- Changing the energy balance.

The foregoing analysis brings us to the conclusion that the development of society population-wise, production-wise and consumption-wise, will have negative effects which might result in irreversible changes and depletion of resources in biosphere, atmosphere hydrosphere and lithosphere creating fatal ecological crisis. That man is facing such an ecological crisis is being felt in an increasing level over the world today. The manifestations of this crisis are many and varied.

Samuel Rayan sums it up thus: "The Earth system is being polluted and destroyed through wasteful, profligate and predatory practices by modern profit-oriented scientific-technological culture, be it industrial, agricultural or communicational. The Earth's standing meaning and history as the Home of Life are under threat of death. Signs of the crisis are many. Precious resources are diverted to destructive and poisonous purposes like arms race, stocking of nuclear weapons, star wars and similar insanities which can wreck and ruin, which have wrecked and ruined the Earth and life partially, totally, irreparably. Nuclear radioactive and other fatally toxic wastes are being dumped in the oceans or on islands and continents of the poor and the powerless. These, together with polluting affluences and emissions from factories, and the acid rains they spawn, are causing large-scale disease and death to plants, birds, beasts, fish, lakes, rivers and human beings. The atmosphere of most cities thick with toxic fumes has become a major menace to health and life....Chemical fertilizers and pesticides poison and weaken the soil, paving the way to greater use of them and greater disasters for the health of the Earth and living things. Water is overused or wasted and underground water sources are being drained dry. Accumulation of Carbon dioxide in the atmosphere and the felling of tropical trees by the thousands to pamper to a consumerist and throw-away culture lead to rise in the Earth's temperature, drastic changes in climatic condition, and depletion of the Ozone layer. These are already threatening to reduce the Earth's luxuriant forests and fertile fields into deserts."[1]

Man today, is seriously warned that if things do not change for better and that if man does not take care of this planet now, one day life on this planet will be miserable and eventually even impossible. So there is an urgency to address the problems related to eco-crisis. That is why the different issues and problems connected with ecology have become a very hot and passionate theme of discussion and debate, not only among academicians and scientists but also among theologians and religious leaders of various religions. Have the Religions to do anything with this problem of eco-crisis? Yes, they do very much. The very root meaning of the word "Religion" comes from the Latin root words, 'Re + ligare' which means 'to bind together' which means in other words, binding together of everything in this universe-humans, plants, animals, earth and environment- in a harmonious way and thus leading to peace and happiness which are nothing but the experience of 'mukti' or 'redemption'. So, every religion, intrinsically and ontologically has a bounden duty to preserve ecology.

Unfortunately, some religious traditions concentrate only on the spiritual dimensions of the followers and preach and teach the ways and means for their eschatological good and happiness in other world. This material world where one lives and moves and which is the locus where one has to workout the ultimate liberation or salvation is often regarded as of lesser value and worth and deserved only second rate attention.

But in fact, a deep concern for Ecology is one of the leitmotifs of all religions. Let us look into the sacred scriptures of these religions, into the teachings of their founders and early prophets, sages and gurus; they all unequivocally admonish people to love and respect the environment and maintain a friendly and caring relation with nature. We should remember that the scriptures of all the major world religions were composed and preached in a distant past and were immediately addressed to peoples who were inextricably bound to the earth for their sustenance and survival and whose social, cultural, and religious customs and practices, feasts and celebrations were inseparably linked with and deeply rooted in nature. It is only natural, therefore that the religious scriptures exhorted the faithful to be gentle and kind, sweet and civilized to the mother earth, which is the source of all life, the matrix of growth and the promoter to prosperity.

No doubt, religions in general are working towards the betterment of the environment. This is particularly true of our Indian religions like Hinduism, Jainism and Buddhism. The Ecological crisis has necessitated a radical rethinking of Christian theology and similarly most of other religions too. Today, it is encouraging to see that all religions in one way or the other are working towards preserving Ecology. Writing on the contribution of all religions toward preserving Ecology will be a lengthy exercise and so, let me limit my treatment to three major religions: namely Christianity, Hinduism and Islam.

Response of Christianity to Preserve Ecology[2]

Often it is believed that Christianity preaches dualism, which separates man from nature, leading to his indiscriminate exploitation of it. It may appear so in a narrow sense. But taking Christianity in a larger context, we can learn that God of the Bible is concerned with totality of human existence, the entirety of human history and with the whole universe. One has to be cautious of what approach is applied in scrutinizing the Biblical passages. Today the Biblical scholars instruct us that applying not fundamentalist approach but contextual approach enlightens us and does justice in getting the fullest message of the Biblical words.

Anthropocentric and Despotic Understandings of the Biblical Texts

In the Biblical language, Man is in 'the image of God'. There is no questioning of man's superiority. But the way he exercises his dominion and the manner in which he deals with material goods, can indicate his basic attitude and disposition towards nature. Man is an intelligent and rational being. His dominion naturally comes from his capacity to think and to reason. He is perfectly within his rights to do so and his dominion is his right as long as it is not abused to serve his whims and fancies without respecting the ecological harmonies in nature.

Though God created both man and nature, man has ignored the role nature plays in his fulfillment. Nature has been desacralized. It is true that nature is not divine. However, to consider it as a mere backdrop for human play is a negative attitude of man. The foundation of this desacralization, we can trace back to the theological creed of Genesis, 1:1-2:4, and the prologue to the Old Testament. This creed affirms that Yahweh, Israel's God is Lord over all, gives all things their order, meaning and value and is faithful to what he creates.

This text coming from the priestly tradition narrates the cosmological history as understood by the people. This Genesis creed focuses on the absolute transcendence of God. Creation by 'word' marks the radical distinction between God and what it creates. In bringing out this distinction and transcendence of God, it is shown that creatures are in no way divine. However, creation is good because it is in a covenant with Yahweh. In the creation of man, emphasis comes on these words, 'let us make men'. Here the speciality in the creation of man is indicated by the use of *'bara'* three times. Otherwise, man too is put in the rank of creatures. Man too is not divine, he is desacralised. But man is created in the *'image of God'*. Besides, he is given dominion over all creation.

The highest point of dominion over nature and the unreasonable manipulation over it came through modern Science and Technology which man applied in his pursuits and with which he has built up today a technological culture. This means an art of living neither oriented for man's global betterment nor channeled for the preservation of nature. Nature has been reduced to merely being, an object for the use of man.

A mechanical view of nature brings about a further rift in man-nature relationship. Descartes and Newton and western thought in general, have contributed much to such a mechanical view. Also, the Orthodox Christian Theology, which has developed arrogance towards nature boosted by Modern Science, has given man wrong attitude, which is at the root of man's ecological troubles. But the Biblical vision is in striking opposition to modern man's attitude.

A proper understanding of this theme is necessary to see if the allegation that the Judeo-Christian theological tradition is responsible for the ecological crisis. The ecological crisis caused by man, the maker-manipulator, forces a theology to reflect on the meaning of the dominion God entrusted to man in his design of creation. God blessed them and said to them, "Be fruitful and increase, fill the earth and subdue it" (Gn.1:28). The dominion entrusted to the privileged creature, must be seen as intimately connected with the dignity of a person, which has been created in the image and likeness of God (Gn.1:27). God, who blesses those made to His image, entrusts them with the Garden of Eden "to till it and to care for it" (Gn. 2:1). Man's conduct should reflect the caring love of God in the wonderful world entrusted to him. Those who refuse to honour God as God and to honour each other as His image, lead to a catastrophe both for man and nature. Their dominion over the world is no longer beneficial, as it should have been according to God's blessing.

The Biblical Perspective[3]

Now let us enter into analyzing verses of the Bible for their proper understanding. The biblical approach to the environmental issue is to ask this basic question: to whom does the earth belong? The first answer is straightforward. It is given in Ps. 24:1, "The earth is the Lord's and everything in it." God is its Creator, and so by right of creation is its owner. But this is only a partial answer. Here is Ps. 11:16: "The highest heavens belong to the Lord: but the earth he has given to man." So, then the balanced biblical answer to our question is that the earth belongs to both God and man-to God because He made it, to us because He has given it to us. This double truth-that the earth is both His and ours-is spelled out more fully in Gen. 1 & 2 and in several verses of Genesis 1 and we may legitimately make three affirmations from this biblical material.

God Gave Man Dominion Over The Earth

We note the two divine resolves of verse 26, "Let us make man in our image and let them have dominion over the earth". We also note the two divine actions in which his resolves were expressed. "So God created man in his own image and God said to them. Fill the earth and subdue it". Thus from the beginning human beings have been endowed with a double uniqueness: We bear the image of God and we exercise dominion over the earth and its creatures. Indeed, our unique dominion over the earth is due to our unique relation with God. God arranged an order, even a hierarchy of Creations. He set man between Himself as Creator and the rest of the creation, animate and inanimate. In some ways, we are one with the rest of nature being a part of it and having the status of a creature. In other ways, we are distinct from nature, being created in God's image and being given dominion.

Biologically, we are like the animals. For example, we breathe like them (living being Gen. 1:22-24, and 2: 7), eat like them (vv.29, 30) and reproduce like them (be fruitful and increase. vv. 22, 28), but we also enjoy a higher level of experience, in which we are unlike the animals and like God. We are able to think, procreate, love, pray, and exercise dominion. This is our intermediate position between God and nature, between creator and the rest of his creation. We combine dependence on God with dominion over the earth.

Man has obeyed God's command to fill the earth and subdue it. This he has done throughout the human history, and today man is obeying God and exercising his God given dominion through his researches, discoveries and inventions in biology, chemistry, physics, and other sphere and in all the triumphs of his technology. There is no question of man having behaved like Prometheus who store fire from the gods. In his progressive control of the earth, man has not been invading God's private sphere and wrestling power from him, not dispensing with Him. It is foolish to draw these deductions. In all his research and resourcefulness, he has been exercising only the dominion God gave him. Developing tools and technology, farming the land, digging for minerals, extracting fuels, damming, rivers for hydro-electric power, harnessing atomic energy -all are in fulfillment of god's command.

In Gen 1: 28 we read "Be fruitful, multiply, fill the earth". This too comes under attack as it is alleged to have fostered the growth and enslaving mentality among the people. In the context of Jewish blessing, only we can understand the meaning of this blessing. It was not so much a command of God to all people, but it applied only to a particular period of history. After the flood, the human family shrunk into smaller families, and was badly in need of God's blessing. The families had to grow from the single ancestor, like Abraham, Isaac, and Jacob and would finally become a big nation. The technique used by the tradition for this purpose is God's special intervention. There is a promise and fulfillment. God promises with a blessing and fulfils it with a special intervention. Therefore, a blessing at the down of creation in this sense could not be regarded as valid for all men at all times.

Our Dominion is a Co-Operative Dominion

In exercising our God given dominion, we are not creating the processes of nature, but cooperating with them. It is clear from Genesis 1 that the earth was made fruitful before man was told to till and subdue it. It is true that man can make the earth more fruitful through many of his innovative and creative efforts. But whatever may be His creative and progressive activities, he is merely cooperating with the laws of fruitfulness, which God has already established. We need to be clear about this distinction that what God gives is "nature" and what we do with it is "culture" or "cultivation". True God has humbled himself to seek our cooperation. But we must also humble ourselves to acknowledge that our dominion over nature would be entirely fruitless if God had not made the earth fruitful, and if He did not continue to "give the increase".

The biblical dominion of man is no despotism. If Genesis authorizes mankind to rule nature, it authorizes only the kind of rule compatible with the Hebrew concept of monarch and, though the Hebrews were aware of other nations having absolute monarchs, their own kings were never so regarded. Rather they were considered answerable to God for the well being of the realm and if they failed in their responsibilities, God would send a prophet to appoint another. The attitude ascribed to David at Chronicles 29:11-14 epitomizes that power to a king. David there attributes all power to God and acknowledges before God that he and his people cannot even offer up gifts which have not first, like everything else, been bestowed through God's grace.

Genesis 2:15 gives a more direct argument for the view that man's dominion was interpreted as the responsible exercise of a circumscribed trust or mandate and not as despotism. The Yahwist account of the creation the Lord God took the man, and put him into the garden to dress it and keep it. "Dress" means to "till for pleasure and profit", and "keep" means "protect from harm". Thus, man is put into the world to look after it for God, and to preserve it as a source of pleasure, and not only as good for good. (Gen 2:9). From this short explanation, it is evident that the Old Testament cannot be reconciled with either the anthropocentric view that everything was made for mankind or the despotic view that people are free to treat nature and non-human creatures as they please.

New Testament Attitude

In the New Testament of the Bible Jesus upheld the Old Testament tradition when he taught that God cares for sparrows (Mt 10:29, Luke 12:6) and 'of the lilies of the field' (Mt 7:28-30). The passage about the lilies suggests that God takes delight even in plants and that value is an independent one. Earlier in (Mk 2) Jesus spent 40 days in the wilderness in the company of wild beasts. Several times, he asked his disciples to sail to the other side of the lake of Galilee to have solitude and escape the multitude (Mk 4:35, 6:45). These passages suggest that Jesus regarded nature not only as resource but also as haven and a source of renewal and he does appreciate natural beauty as we find in the passages about the lilies. In Rom 8:21 St. Paul held that every creature was in travail awaiting release from decay and participates in the liberty of the children of God. According to St. John the tree of life (Rev. 22:2) symbolizes final restoration of the original tree of life of Genesis (2:9) and with it a restoration of the original Garden. These and other passages altogether prelude despotic and anthropocentric reading of the New Testament, with regard to nature.

Our Dominion is a Delegated and a Responsible Dominion

Our dominion does not belong to us by right but only by favour. The earth belongs to us not because we made or own it, but because its maker has entrusted its care to us. God's continuing ownership and caring supervision of the earth is asserted many times through scripture (Refer Ps 24:1, Ps 50:10, 11, Mt 5:45, Mt 26: 38, 30). If therefore God has delegated dominion over the earth to us, with a view to our cooperating with Him and sharing its produce with others then we are accountable to Him for our stewardship. We have no liberty to do what we like with our natural environment; it is not ours to treat as we please, and 'dominion' is not a synonym for 'destruction'. Since we hold it in trust, we have to manage it responsibly and productively for the sake of both our own and subsequent generations.

The Bible thus is not despotic to nature and the Judeo-Christian traditions has stressed the responsibility for nature and offer us resources, which may, in refurbished form, allow us to cope with the modern environmental problems. The Jewish-Christian doctrine of creation is, therefore much less responsible for the ecological crisis than is suggested by its critics. On the contrary, the biblical foundations of the doctrine would tend to the opposite direction away from a license to exploit, towards a duty to respect and protect.

Love of nature is the apex of ecological responsibility. It is based on the will of the creator who created it and saw that it was good. The goodness of creation reaches its full meaning in the words of Jesus Christ. "God so loved the world that he gave his only son" (Jn 3:16). Our love of God itself requires a love for nature. To love God with all our heart, with all our strength is to be interested in the well being of His creation. God placed the universe at the service of love. In its integrity nature is disposed for the good of all human being, all animals, insects and plants and the ecosystem. Each draws its own nourishment, and contributes to the further benefit of others. Ecological responsibility then is the central act of justice and love for God and for our fellowmen on earth. A world that is ecologically managed is more valuable to humanity and creatures. To love one's neighbor is not only to give him bread and water, but also to keep his air, water and food free from chemical poison. Well-being of man and his neighbor is intimately bound up with the well-being of nature. Exploitation of nature and selfishness cannot go with love of God who wants us to respect the ecological order.

Moral life of man is very much bound up with his natural surroundings. Pope Paul VI in an address to the people said: "We cannot, in fact, remain indifferent to the world wide anxiety aroused by pollution of these natural elements to which man's physical and moral life is inevitably bound. Whereas a wholesome natural environment actually helps him to rise towards the divine world, polluted world can handicap him in his effort to attain God."[4] So the fullness of mutual life with God and with fellow men can be developed only in an ecologically responsible life.

Response of Hinduism to Preserve Ecology[5]

What are the essential teachings of Hinduism about Ecology? According to Hinduism all living beings are sacred because they are parts of God and should be treated with respect and compassion. This is because the soul can be reincarnated into any form of life. Hinduism is full of life stories that treat animals as divine, such as how Krishna used to herd cows, or how Hanuman was a faithful servant of Shri Ram. Most Hindus are vegetarian because of this belief in the sanctity of life. Even trees, rivers and mountains are believed to have souls, and should be honored and cared for.

The virtue of a simple life has always been prized in Hindu society. Teachers or Brahmanas, are advised to live on the charity of others and not accumulate too much wealth. The most highly respected person in Hindu society is the Sadhu, or Sage who lives outside normal society, in forests or caves or travel on foot from one town to another. Sadhus take pride in living simply and consuming as little as possible. Thus Hinduism stresses that true happiness comes from within not from outer possessions. This means that the search for material possessions, and the consumption of materials and energy it brings, should not be allowed to dominate life. Life's main purpose is to discover the spiritual nature and the peace and fulfillment it brings. The efforts to exploit the things of this world are considered by Hindu teachers to be a distraction from this central purpose of life. Many eminent thinkers of Hinduism have elaborated their views on ecology.

Swami Vibudhesha Teertha and Ecology[6]

According to Hindu religion, "dharanath dharma ucyate"-that which sustains all species of life and helps to maintain harmonious relationship among them is dharma. That which disturbs such ecology is adharma. Hindu religion thus wants its followers to live a simple life. It does not allow people to go on increasing their material wants. People are meant to learn to enjoy spiritual happiness, so that to derive a sense of satisfaction and fulfillment they need not run after material pleasures and disturb nature's checks and balances. They have to milk a cow and enjoy, not cut at the udder of the cow with greed to enjoy what is not available in the natural course. Do not use anything belonging to nature such as oil, coal, or forest, at a greater rate than you can replenish it. For example, do not destroy birds, fish, earthworms, and even bacteria which play vital ecological roles and once they are annihilated you cannot recreate them. Thus, only can you avoid becoming bankrupt, and the life cycle can continue for a long, long time.

The message of the Bhagavad-Gita is conservation and protection of ecology. Life is sustained by different kinds of food; rainfall produces food; timely movement of clouds brings rains; to get the clouds moving on time *yajna*, religious sacrifice, helps; *yajna* is performed through rituals; those actions that produce rituals belong only to God; God is revealed by the Vedas; the Vedas are preserved by the human mind; and the human mind is nourished by food. This is the cycle that helps the existence of all forms of life on this globe. One, who does not contribute to the maintenance of all this cycle, is considered as a destroyer of all life here. Hinduism is also a religion that is very near to nature. It asks its followers to see God in every object in the Universe. Worship of God in air, water, fire, sun, moon, stars, and Earth is specially recommended. Earth is worshiped as the spouse of God, hence very dear and near to God. All lives on earth are considered as children of God and Earth.

There is no life that is inferior. All lives enjoy the same importance in the Universe and all play their fixed roles. They are to function together and no link in the chain is to be lost. If some link is lost, the whole ecological balance would be disturbed. All kinds of life-insects, birds, and animals-contribute toward the maintenance of ecological balance, but what is humanity's contribution toward this? Humans are intelligent animals; therefore, our contribution should be the biggest in maintaining the ecological balance of the whole creation. *Why was humanity created?* There should be a purpose for the creation of humanity. What it might be! We could be the sustainer of interrelationship among numerous life species on earth. We are ones who can see God, and all objects, as the controller and sustainer of ecological balance. All other animals play their roles without knowing what they are doing, but humans do everything with full consciousness. God created the human mind to see His own reflection as in a mirror. The human mind can mediate on God and know him more and more.

Dr. Sheshagiri Rao and Ecology[7]

According to Dr. Sheshagiri sacrifice does not just mean ritual worship–it means an act that protects life. Personal health depends on eyes, ears, and other sense organs working together in harmony; human prosperity and happiness depend on a well ordered society and nature; the universe is sustained by the cosmic powers such as the sun and moon working together in unison. Sacrifice reinvigorates the powers that sustain the world by securing cosmic stability and social order. It activates the positive forces of the universe, and protects the Earth from degeneration. God's creation is sacred. Humanity does not have the right to destroy what it cannot create. Humans have to realize the interconnectedness of living entities and emphasize the idea of moral responsibility to oneself, one's society and the world as a whole. In our cosmic journey, we are involved in countless cycles of births and deaths. Life progresses into a higher form of life or regresses into lower forms of life based upon our good action or *bad karma*. Kinship exists between all forms of life. Reincarnation warns us against treating lower forms of life with cruelty.

The principle of cow protection symbolizes human responsibility to the subhuman world. The cow serves humans throughout and even after death. The milk of the cow runs in our blood. Its contributions to the welfare of the family and the community are countless. Hindus pray daily for the welfare of cows. When the cows are cared for, the world at all levels will find happiness and peace. Hindus also revere the earth as mother. She feeds, shelters, and clothes us. Without her we cannot survive. If we as children do not take care of her we diminish her ability to take care of us. Unfortunately the Earth herself is now being undermined by our scientific and industrial achievements.

Shrivasta Goswai and Ecology[8]

According to Hindu philosophy, the goal of human life is the realization of the state of peace. Dharma, loosely translated as religion, is the source by which peace can be fully realized. This peace is not the stillness of death; it is a dynamic harmony among all the diverse facts of life. Humanity as part of the natural world, can contribute through dharma to this natural harmony. Just as the human child has to be nourished by Mother Nature, so, the human spirit has to be embraced and loved by beautiful nature. So, the human being who has grown old or sick has to be supported by caring nature. If humans distress the mother, rape the beauty and beat the caring nurse, what happens? The relationship collapses and the family will be broken.

If we think of the environment as our home and all of its members as our family it is clear that the key to conserving nature is devotion, love –

giving and serving. Nature (*Prakriti)* as the feminine can give and serve. But the role of humanity (*Purshua)* is then to protect. Nowadays (*Purshua)* humanity is interested not in protecting but in exploiting, so Prakriti (nature) has to defend her. This is why we see nature in her furies manifestation - in drought, floods or hurricanes.

Response of Islam to Preserve Ecology[9]

Islam (meaning "peace") is a world religion, which teaches about preserving ecology in clear terms. Through its scripture *Holy Quran* it pointed out that man is only a part of the nature's web and he has no right to disturb the web. In case he does, he will have to face harsh punishment on the Day of Judgment and lose the right of trusteeship of nature. Islam offers ten injunctions to preserve ecology and they are the following:

His role as *Khalifa* (stewardship). Muslims believe that Allah created humans to be guardians of his creation. In other words, nature does not belong to us to do with as we wish, but is entrusted by Allah to our safekeeping. "The world is green and beautiful, and Allah has appointed you his guardian over it", taught the prophet Muhammad. As *Taskheer* (one who complies) he has to comply and manage the earth purely in accordance with the will of Allah, the supreme creator. *Jamal* (aesthetic) is the manifestation of beauty in all the creatures of God: the women, the birds, the flower, the snow-cad mountains, the rainbow in the sky–all these are the solid proofs of God's love for beauty and no one on this earth is capable of excelling the master painter (God). Therefore, man has to take adequate care for the preservation of the bounties of beauty lay scattered in the creatures of God.

Amanah (trusteeship) stands for the trusteeship devolved on men. Like an honest treasurer who keeps wealth and money in his custody, he has to preserve and keep intact the beauties of God...resplendent all over his creation and if he uses it, he has to use it with caution and that too for common good. Closely connected with this, there is another word found in Holy Quran. That word is *Krah,* which means accountability. *Raham* (mercy) is one of the greatest attributes of God. God created nature for the benefits of human beings as an expression of Raham and so, he allowed men to use fruits and vegetables and to take some service from the animals–the camels, horses and ponies but there should be no excess, everything should be used in moderation–be it the use of fruits or food grains or the flesh of animals and other things of nature falling under this category. As Allah grants all comforts even to disbelievers because of His Raham, he wants the human being to reciprocate this Raham towards all his creations.

Human beings have been strictly directed by the Holy Quran to observe *Adal* (Justice) in all actions. If he is a keeper of animals, he should feed them properly and while loading camels, horses and bullocks, he should keep in mind the health of the animal concerned. Indeed the SPCA

(Society for the Prevention of Cruelty against Animals) should take cues from this injunction of the Holy Quran to protect the lives of ailing and half-fed animals. The provisions of water and fodder must be adequate for them. If a man does not observe this norm, he will be declared a sinner. Indeed, his disposition towards nature should be like a father towards his child.

Quran significantly mentions the word *Mizan* (balance) which rightly shows the weight of a commodity and God has warned that *Mizan* or balance should be accurate because he likes only those persons who observe the principle of balance which is clearly manifest in God's creation, i.e., the nature.

The Holy Quran ordains that a person should be *Zarar* (non-injury) and harmless towards nature i.e., he should not injure plants, animals or other creatures of God without valid reasons. Snake should be left unharmed if it does not attack him (man), he should not kill a lion or even the beasts for his pleasure because all these animals have to play their assigned role in nature as ordained by God. The *Zarar* explains that men should not do anything injurious to nature for it will go against the wishes of Allah. The Quran asserts, "created beings are the dependents of Allah and the creature dearest unto Allah is he who does most good to God's dependence.

Almighty God has set limits in mutual dealing not only with fellow human beings but also with other creatures of the world created by God. The limit should not be crossed otherwise, it will tantamount to transgression. Only God knows what the limits in all walks of life are. Itgha (non-transgression) explains that human beings should act strictly within his limits and need not transgress authority vested in him by God; the Sufi poet Rahim also sang the virtues of moderation and the avoidance of excesses. *Fasad* (corruption) includes not only fighting and rioting among themselves but also includes destruction of nature by the use of lethal weapons like atom bomb, which destroyed the entire flora and fauna of the regions as happened in Hiroshima and Nagasaki and more recently in Iraq. Fasad emerges out of transgression or overdoing. Man is called to avoid any form of corruption.

Action Programmes

The famous words of the Greek Philosopher, Socrates, are these: "Know thy self". Likewise the Prophet Jeremiah says in the Bible, "The Laws of the Lord are written in your hearts". Kant also holds similar position when he speaks about the Moral Categorical Imperatives. Along with these wise sayings can be said that a majority of the humans in this advanced world clearly know what to do and what not to do, both at the personal level and at the social level on the serious issue of preserving Ecology. Let all discussions stop and acting start! Let every one take personal interest in preventing activities, which endanger ecological order, and promoting eco-spiritual activities.

- Religious stories, legends and myths are a rich mine of ecological wisdom. These can be collected and culled and disseminated to both for self-education and others by way of telling stories and preaching, showing pictures and films and workshops. This should be done constantly.
- Let us reevaluate our concepts of the Divine in such a way that unwanted dualism and dichotomies between the supernatural and the natural can be reduced if not completely minimized.
- It is time to start a fresh reflection based on Earth wisdom or Earth rationality in order to articulate a new Epistemology of interconnectedness and Ontology of relatedness and formulate a religion at the intersection of these two.
- A discovery can be done of the ethical and beautiful based on an organismic view of the Cosmos and the human community.
- It is time for building Eco-movements based on Eco-commonalities from various religions emphasizing the need for sustainable practices in economy, agriculture and tourism.
- Let us make use of the information technology and communication media in emphasizing eco-values of green and clean earth and more accountable sharing of the resources as simpler modes of the living.
- In celebrating Christian religious festivals like Christmas, Palm Sunday, Easter, Festivals of Saints and Parish Festivals the believers may be instructed to reduce using leaves, plants and trees for the purpose of decorations and performing religious rituals though as symbols they have religious meanings.
- Firecrackers and loud speakers are used in almost all secular and religious functions in our societies. But firecrackers with high magnitude and loud speakers with high volume create air and noise pollutions. It is sad that huge amount of money is spent on such items. "Use but not abuse; avoid excess leading to waste", is an important teaching of Islam. Worshippers and festival celebrants may be instructed by secular leaders to minimize or even stop using the eco-inimical items.
- Comparing with the recent past, today public celebrations have been increased both at secular and religious levels. Witnessing the way in which celebrations are carried out today, one can speak for sure that hardly there is any celebration, which is conducted without disturbing the healthy and peaceful atmosphere of the environment. Disorders to nature are brought out more through religious celebrations than the secular ones. So awareness should be created among people

regarding the need for reducing public celebrations, particularly the religious ones.

- Christmas celebration is an important one in Christianity. The crib-decoration is the center of it. Christmas crib-competition is an enthusiastic one held in many centers of Christmas celebrations. People can be instructed to reduce the use of nature-elements like leaves, plants and other vegetations, and instead to make use of waste things: "Art from waste" principle can be followed.
- Exchange of greeting cards is part of almost all our celebrations. It may be meaningful to the exchangers but it is done at the cost of tampering with nature. We know that trees are used for manufacturing greeting cards. Efforts should be taken to simplify or to reduce the use of greeting cards so that cut and use of trees can be minimized. Other alternatives can be followed to exchange our festal greetings. Similarly, efforts should be taken to reduce the use of papers for decorations in celebrations.
- Also, a large number of plastic cups, plates, garlands and decoration articles are used in religious functions. It has been estimated that a plastic material takes around eighty years to decompose. So, ways and means of the use of plastic articles should be seriously sought for.
- It is a common sight that spacious and grand buildings for worship such as temples, churches, mosques and gurudwaras are coming up everyday in great numbers, consuming a lot of wood works for doors, windows and other wooden articles. Now-a-days more wood-works are introduced in all types of stately buildings for elegance and it is easy to cut carvings in wood works to make it more architectural. Moreover, huge amount of money is being spent for constructing religious buildings, which would in turn affect economically the human community. Buildings of worship could be simplified and be built not at the cost of endangering nurturing nature.
- We need to critically rethink certain cultic functions. In general people, carried by varied superstitious beliefs use (rather abuse) a lot of stationary for religious purposes. For instance, in the name of offerings, people carry out some cultic functions like, "breaking 100 coconuts, garlands made by limes, burning 1000 candles, and pouring large quantity of water on wheel of cars during processions. Utility of all such activities are controversial and debatable one. We need to critically approach these issues from an eco-friendly and eco-spiritual perspective.
- Religion no doubt, is good but common man is intoxicated with religious feeling and sentiments so much that he is ready to spend (or waste) any amount of money for any sort of religious celebration, and

when it comes to the question of spending money for necessary items like education of their children, small saving scheme, in view of the future, improving his livelihood in future stingy etc. he becomes very skeptical and negative. Through religious instructions, people should be instructed on value of money and the proper use of it, which moved in turn, will pave the way for minimizing the destruction of nature.

- Trees can be planted and clean environment can be created around the places of worship. The centers for worship can allot certain amount of money from its income on a fixed basis for Eco-awareness programmes and Eco-preservation activities. It would facilitate the religious atmosphere. Nature thus can create conducive atmosphere to raise our minds and hearts to God.
- Car processions during religious celebrations normally bring inconveniences to the common man like traffic jams and diversions of travel routes. Care should be taken to check unhygienic practices during religious functions like improper sanitation facilities, selling eatable in an unhealthy atmosphere, dirtying public places with all sorts of remains etc.
- Taking stock from respective Scriptures and Holy traditions we should earnestly involve ourselves in protecting and enhancing the environment and unequivocally condemn every act of abuse and violence against Nature. The serious minded religious of all faiths should interact not only within their own faiths but also with the eco-activities of other faiths and non-believers to save and safeguard the environment with courage, devotion and commitment.

Conclusion

There is a growing awareness among people all over the world today that we are facing an ecological crisis, which is so serious that the Earth's standing, meaning and history on the Home of Life are under threat of death. That is why the issues and problems of Ecology have become a very hot and passionate theme of study not only among academicians and scientists but also among the religious leaders of various religions. The valuable contributions to Ecology by the three major religions, namely Christianity, Hinduism and Islam are stated in this paper so that these sound thoughts and deep awareness may force human race to actions. Going through the sacred scriptures and holy traditions of various religions we have received enough and more thoughts concerning the Ecological problems. Remaining in this level is of no use unless we are led on to concrete actions. This paper does not claim that it exhausts in discussing all matters of religions to preserve Ecology. However it surely shows the way to the readers towards saving and safeguarding the environment.

Notes

1. Quote taken: Joseph Mattam and Jacob Kavunkal, ed., "Ecological Concerns" Seva Sadan Training Institute, Bangalore, 1998, p. 11.
2. Kurien Kalapurachal, Ecology and Religion,: the Christian perspective, Rajdeva Narayan Janardan Kumar, ed., *Ecology & Religion,* Deep & Deep Publications Pvt. Ltd., New Delhi, 2003, pp. 101-105.
3. Quote taken: ibid. Pp.122-123.
4. The quotations from the Bible used in the article have been mainly taken from *The Holy Bible*, The Theological Publications In India, Bangalore, 1973.
5. Internet article: "Faith and Ecology" taken from the websit, <htt://www.arcworld.org/faitha.asp? pageID=69>
6. Acharya of Madhvacarya Vaishnavas,Udupi,and Central Advisory Committee Member of the Visva Hindu Parishad.
7. Chief Editor of the Encyclopedia of Hinduism.
8. Vaishnava Acharya of Shri Radharaman Temple, and Vrindavan Chairman of the Vrindavan Conservation Project.
9. Janardan Kumar, "Islam and Ecology" Rajdeva Narayan Janardan Kumar, ed., *Ecology & Religion,* Deep & Deep Publications Pvt. Ltd., New Delhi, 2003, pp. 143-161.

Bibliography

Narayan, Rajdeva and Kumar, Janardan, ed. (1998), *Ecological Concerns,* Seva Sadan Training Institute, Bangalore.

Mattam, Joseph and Kavunkal, Jacob, ed. (1998), *Ecological Concerns,* Seva Sadan Training Institute, Bangalore.

Thottakara, Augustine, ed. (2000), *Eco-Dynamics of Religion: Thoughts For The Third Millennium,* Journal of Dharma & Dharmaram Publications, Bangalore.

Chetti, D. Daniel, ed. (1991), *Ecology and Development,* Uelci/Gurukul & BTE/SSC, Madras.

Banerjee, Suresh Chandra (1980), *Flora and fauna in Sanskrit Literature,* Naya Prakashan, Calcutta.

Biardean, Madelline (1989), *Hinduism: the anthropology of civilization*, Oxford University Press, Delhi.

Dweived, O.P., *Environment Crisis And Hindu Religion*, 1987, Gitanjali Publishing House, Delhi.

Agwan, A.R. (1992), *the environmental concern of Islam,* institute of objective studies, New Delhi.

Ba Kader, Abu Dark Ahmed (1995), *Environmental Protection In Islam,* Island Press, Washington, D.C.

Blackstone, W.T. (1974), Philosophy And Environment Crisis, Athens, University of Georgia Press.

Carmody John (1983), *Ecology and Religion: Toward An New Christian Theology of Nature,* Paulist Press, New York.

Fernandez, Archbishop Angelo (1991), *the Ecological Crisis and the Search for Peace,* World Conference on Religion and Peace, New Delhi.

Internet Article, "*Faith and Ecology*" taken from the website, <htt: / / www.arcworld. org/ faitha.asp? Page ID=69>

NATURE PROCLAIMS THE GLORY OF GOD

Priyanka Parmar & Deepa Pal

"When wisely trod, the path to God through Nature employs every faculty inherent in man." ***–Flower A. Newhouse***

"You can't prove God exists and you can't prove God doesn't exist." This is the response one often hears when the question of God's existence is raised. We know that all the things that we see around like trees, mountains, seasons, storms, and thunders all are collectively called nature. Does nature give evidence to the existence of God? Through history she has been subpoenaed for many moral and spiritual arguments. But perhaps she speaks only for herself, in finest poetry. The natural benefits that we enjoy or the natural calamity that we face day to day are neither in the hands of man nor in the hands of any other living being present on the earth. It in itself gives us the proof that some super natural power exists which has the power of controlling the ultimate destiny and life of mankind.

The popular media have recently pointed toward an apparent convergence between science and religion. Modern Particle Physics appears to give credence to New Age explorations of quantum mysticism. Meanwhile, certain aspects of Big Bang cosmology are supposed to indicate the intervention of a supernatural intelligence. Some scientists identify so-called anthropic coincidences, aspects of our universe which make life possible and seem so remarkable that they must have been arranged intentionally. However, this should not be seen as a widespread movement within the scientific community. As always, some scientists are believers who have interpreted new data as supporting conventional ideas. Their statements are debatable, aren't affirmed by the mainstream, and should not be viewed as a consensus.

On the contrary, as many researchers as ever assert that the universe can be explained without the need for divine intervention. In looking for a supernatural force, one would predict aberrations from natural laws. But despite the evidence we might hope to find, the net balance of energy in the cosmos appears to be near zero. Scientists have yet to detect any input at the point of origin or anywhere else. There is no clear fingerprint of God. The Big Bang was entirely within the realm of natural possibility. Life on Earth is a marvelous cosmic circumstance - a rarity perhaps, and fortunate from our perspective but in such a vast sea of possibility, rarities occur.

One writer who has devoted much of his life work to balanced skepticism is Victor J. Stenger, Professor Emeritus of Physics and Astronomy at University of Hawaii. The sample writings on his website do not attempt to disprove the existence of God, but simply to refute unfounded claims. On the other extreme are those who insist that God does not exist. The difficulty with metaphysical naturalism, the view that nature is a closed system with no need of explanation, is that it offers what many people will view as a sterile existence devoid of spirit. Faith has given many of us a purpose for living, a structure for compassionate society, and strength to face life's trials.

Environmental Imbalance: Leading to Extinction of Mankind

Today we know that the ice at the glaciers is melting slowly, natural calamities like earthquakes, thunder storms, heavy rainfalls, etc. are occurring rapidly and the sole reason for such things happening today is the imbalance produced in the nature by human beings from last so many years. Man has no control over these natural destructions and he can only wish and pray to the almighty to stop all these processes. If human beings will keep on doing harm to nature, the day is not far when there will not be any existence of mankind on earth. So this gives us enough idea that man need to take immense care of nature and be very careful in dealing with it otherwise, nature will not leave any stone unturned to ruin the human race.

We turn to nature to nurture this faith. Many of us have had our most meaningful spiritual experiences in relationship to the environment. Beauty provokes us to transcend our mundane concerns, shortcomings and conflicts. We have heard the voice of God in our woods and waterways, in bird song or the remarkable delicacy of a flower. We will not readily abandon the idea of a higher power, or ascribe all this complexity to the whim of chaos.

Questions about the existence of God wouldn't be so polarizing if we weren't obsessed with certainty and objectivity. The ancients had to be satisfied with guesswork to figure out how the universe functioned. Anything that didn't make sense could be attributed to supernatural powers. Now we have more tools for finding real answers to questions about nature. But the spirit world eludes our finest-tuned instruments. Probably no telescope will ever reveal the face of God in a nebula or supernova, like a glimpse in a room of mirrors. Outside of our own hearts, we search in vain. Such an approach to understand Nature requires reverence. In contrast to a casual or indifferent attitude, the sensitive individual looks upon Nature as a sacred creation of God that deserves to be valued and revered.

Esoteric Truths

Some of the most intriguing aspects of quantum mechanics, particle physics and chaos theory seem too esoteric for the layman. The situation is almost reminiscent of the difficulties, which have always surrounded mysticism. Throughout the ages and in various cultures, mystics have explored a direct and unusual personal experience of God. This was usually deemed to require special spiritual abilities. An aspiring mystic must learn special techniques for meditation and reflection from a *guru;* otherwise his or her efforts could be misguided and even dangerous. Ultimately, mystics have contributed to all the major belief systems of the world. They tended to bring compassion and tolerance to the most destructive and insufferable religions. But their teachings had to filter down to the masses in a simplified form. Some of the wisdom was kept hidden. This practice was not elitist. It was necessary because not everything would be comprehensible or practical for the average person.

In our individualistic era, everyone wants the best of everything, even knowledge, and we want it quickly. Catering to this, many contemporary brands of mysticism offer little more than a quick fix, a readily marketable product. Most people are too impatient to devote their entire lives to meditation on the path to enlightenment. The same is true of scientific truth. Few are willing to devote years to studying and contemplating the universe with an open mind. But we demand to know, we cherish our preconceived notions, and we want answers fast. The popular media and the pedagogues pander to these expectations. Awash in speculation and hyperbole, we have a hard time recognizing what's bogus. And usually we would rather dismiss any facts that threaten our familiar way of thinking.

Looking for a Teacher

So how can modern science lead us closer to God? It probably can't, if we're seeking concrete spiritual facts to support certain doctrines or cosmology. We will find answers, but the best lesson we can hope to learn from gazing at the skies is one of humility and wonder. For deeper spiritual meaning, and for resolving moral and social problems, we must rely on the spirit.

Nature in our backyards is important because it is our fragment of the cosmos, immediate, real and interactive. It can give us a personal experience of transcendence. We have its energy before our eyes, in our noses when we smell the spring mud, or between our fingers when we whittle a willow branch. It is not always the bucolic image portrayed in lovely albums of wildlife photography. Sometimes cruel and dangerous, it is yet constructive, creative, beautiful and powerful, a fitting embodiment of God. Nature's ways are worthy to inspire us. It leads us into the realm of the imagination, our inner playground and the workshop of enlightenment.

Supernatural Power and Cultural Evolution

An evolutionary perspective that emphasizes the observable consequences of religious faith and recognizes how beliefs can give rise to behavior that regulates access to essential material resources, may be productive in explaining the early origins and universality of belief in supernatural power. All human cultures appear to include faith in supernatural power, and it seems that this tradition has played an active and essential role in influencing how our ancestors perceived and adapted to their environment. This relationship between human belief and adaptive behavior may have played a critical role in shaping cultural evolution.

The history of any society can demonstrate how sacred beliefs change over time, how they are created or borrowed, consciously or unconsciously, and how they may be subsequently abandoned. Beliefs about the natural environment and the supernatural environment are part of a larger system of learned ideas and customs that comprise a culture. And entire cultures change over time in the same way beliefs do. Human populations no longer adapt to environmental change by evolving generically. We now adapt to change by altering our beliefs and behavior. Human evolution from Australopithecines to Cro-Magnon, the first of our ancestors to have the physical features that characterize all human populations today, depended largely on slowly changing gene frequencies. For at least the past 60,000 years, human biological evolution has been relatively inactive. During this time, the evolution of beliefs has come to determine which populations had the greatest capacity to survive.

Some populations maintain customs that threaten the survival of certain individuals and at the same time, help to ensure the survival of the society. For example, the belief that first-born females are cursed and the practice of female infanticide in a tribal society can effectively check population expansion in an environment where unlimited growth would likely lead to mass starvation. Similarly, the traditional prohibition against slaughtering cows in India, where famine was not unusual, meant that many poor people would starve rather than eat beef. But the belief that cows were sacred enabled them and the oxen they produced to survive, thus ensuring that beasts of burden would be available for their essential role in this agrarian society. Had the consumption of beef not been taboo in this largely impoverished and rapidly expanding population, cows would be consumed faster than they could reproduce, oxen would no longer be available for plowing, agricultural output would steadily decline, and the entire society would be hard-pressed to survive.

Proposing that traditional faith in supernatural power has utility for the believers is not at all new. Sigmund Freud, who did not much care for religion, nevertheless gave credit for emotional functions such as reducing anxieties. Emile Durkheim, regarded as the first sociologist, proposed that belief in magic and spiritual beings was highly beneficial for the maintenance of society. But anthropologists now are suggesting that belief in supernatural power is more practical than meeting mere emotional and social needs; it may have played an essential role in ensuring that our ancestors' material needs were met.

God–The best Explanation for the Beginning of the Universe

It is quite logical to argue that whatever begins to exist must have a cause. Now this fact requires no extra support that the *universe* began to exist, therefore, the universe *has* a cause. Whatever begins to exist must have a cause. Most of us have no problem accepting this principle. We assume its truth in virtually every aspect in our daily lives. Our experience always confirms it and never denies it. But surprisingly philosophers have been unable to prove its veracity. Nevertheless, it has always been a fundamental first principle of philosophy and science that *"from nothing, nothing comes"*, *"being cannot come from non-being"*. Even the great skeptic David Hume, who argued that we could not prove the causal principle through ordinary means, still believed it to be true and thought a denial of it was absurd, "I never asserted so absurd a proposition that anything might arise without a cause."

Surely it is more reasonable to hold to this premise than to believe that things pop into existence out of nothing and by nothing. Can we reasonably disagree with the atheist Kai Neilson when he writes, "Suppose you suddenly hear a loud bang and you ask me, 'What made that bang?' and I reply 'Nothing, it just happened! You would not accept that-in fact you would find my reply quite unintelligible." It is quite logical to believe that there is and has always been an extra-ordinarily powerful and intelligent source behind the creation of this entire universe. This source has to have innumerous energy, logic, reasoning and authority as well as capability for giving birth to this universe, nature and life. Can this source be something else or somebody other than the God?

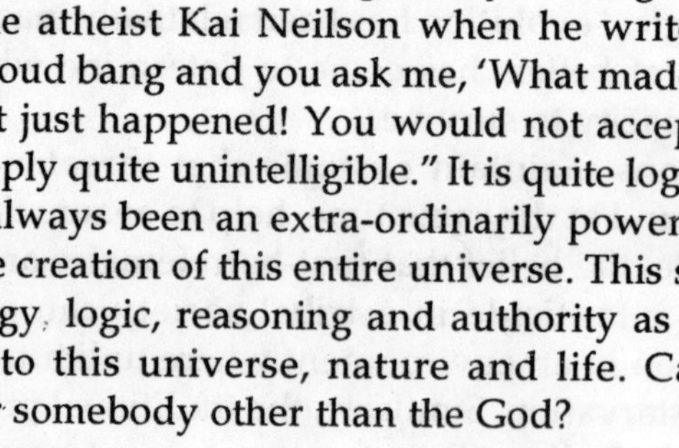

"Now nature is not at variance with art, nor art with nature; they being both the servants of His providence. Art is the perfection of nature. Was the world now as it was the sixth day, there were yet chaos? Nature has made one world, and art another. In brief, all things are artificial; for nature is the art of God" says *Sir Thomas Browne.* "I trust in Nature for the stable laws of beauty and utility. Spring shall plant and Autumn garner to the end of time. I trust in God—the right shall be the right and other than the wrong, while he endures; I trust in my own soul, that can perceive the outward and the inward, Nature's good and God's", says *Robert Browning.*

It is very well said, "When one experiences Nature through the eyes of inner perception, there is nothing common place in this kingdom. God is in every form of creation. Here you can meet the Creator face to face, if anywhere on Earth, yet very few venture into Nature with the purpose of making His acquaintance." Our nature is beautiful and it has been fulfilling, each and every need of human beings. Who else other than God Himself can do that much for his children? We have already talked enough about the existence and non-existence of God. If at all God exists which we are very much in favor of believing, then He lives in nature He is nature. We believe that if one asks for the visible proof of God and we often come across the question that has anyone seen God? Yes, we have and nature is the only visible proof of God's existence. Therefore, nature proclaims the glory of God.

"Nature is like another dimension and we know to approach it with faculties that we do not use perhaps in the same way at other times. Nature, whether it is approached in our gardens or in the wilderness, or on mountain peaks or in the countryside, is now the enlightener. She requires from those who comes to her not the desire to be entertained. If we are to be enlightened in the wilderness, we must not go there just for vacation, just to relax, but we should go there spiritually on tiptoe with all our faculties alert, and we should know that in going to Nature, we are approaching a storehouse filled with treasure, deep potent, spiritual treasures, treasures intended to be released after the very potent season of Eastertide."

God, therefore, created the present world as a perfect place to fulfill His purpose, with all nature under His command. The causality resulting from the quantum nature of matter gives God the power to control events without altering His laws of nature. Therefore, even when God does not miraculously intervene in worldly happenings, as when things occur through the laws of nature, by accident, or as a result of man's free will, all happenings ultimately result from God's will.

Nevertheless, when the destiny of any living creature is linked with that of men, it is also individually judged and guided by God. Therefore, when a snake bites or a bird is captured, it happens only through God's decree. Similarly, the destinies of domestic animals are linked with their owners, as are all their possessions. Indeed, God watches the animals of the righteous to such an extent as to prevent them from doing anything embarrassing to their owners. He is the master of each and every thing happening anywhere and anytime. There may and may not be instantaneous and explicit evidences of His existence but His divine presence can be felt everywhere. Of course, the faith and belief play important roles in developing this insight.

Conclusion

Nature has got the power of making or destroying the lives of all the creatures. The favorable conditions of nature have given rise to life on planet Earth. Why is it that only we have life on this planet and all the other eight planets of our solar system have no life? Somewhere our nature is responsible for it and none other than the almighty God could control nature. There is but one God, He who gives all blessings to man and other creatures. His greatest gift is that He reveals Himself. But in many tangible ways He cares for man and provides for his growth. Man has always been searching for the reason behind his existence. He has been trying to find every possible way to know the origin of his existence. All human beings believe in the existence of the great almighty power in some or the other way. Whether they call him as Rama, Jesus, Pagamber or Guru Nanak they all believe that there is one great power that is above all the others and He has the control on every single thing present in this universe. Now, if we talk about nature none of us actually knows how this nature is continuing on its own pace and it has got so much of diversity in itself that none except God could be the creator of it. It very well in itself explains the existence of God in nature or nature being the visual manifestation of God.

References

David C. Lindberg and Ronald L. Numbers, ed. (1986), *God and Nature*, University of California Press.

Soren Kierkgaard, Papers and Journals: A Selection, Penguin.

S. Morgan, Thaler (1999), *Worship Evangelism: Inciting Un-believers into the Presence of God*, Z Publishing.

SPIRITUAL ECOLOGY: A BLEND OF RELIGION, FAITH AND SCIENCE

Priti Gupta

Ecology is defined as the science of the relationships between organisms and their environment. The term was coined in 1866 by the Darwinist German biologist Ernst Haeckel from the Greek word 'oikos' meaning "house" and 'logos' meaning "science". The environment includes both the abiotic environment-non-living things like climate and geology-and the biotic environment-living things like plants and animals. To be more precise it is the study of the relationships between all living organisms and the environment, especially the totality or pattern of interactions; a view that includes all plant and animal species and their unique contributions to a particular habitat.

Generally we think of these relationships as being of a physical nature but historically and practically this is not true. A being's environment is not only physical but also emotional as well as spiritual. Those inquisitive minds who have been researching the spiritual environment of this world will usually agree that the spiritual precedes the physical. That means that if there is a physical problem there is usually a spiritual problem behind it and if we address the physical without addressing the spiritual the resulting changes will be short lived. If, however, we address the spiritual and set the underlying problem right, the physical will rectify itself. This is because right relationships and honoring the spiritual ensures right relationships and honor on the physical and this is the essence of ecology and nature. As observed by Francis Bacon-We cannot command Nature except by obeying her.

In nature everything is interdependent. Symbiosis or cooperative living together, is far more fundamental than the struggle of the "survival of the fittest." We depend on smaller lives—cells, microbes, atoms—and they need us. This symbiosis works along with pressures of the physical environment which is a factor in the selection of better adapted organisms. Struggle does occur in nature, but harmony prevails. Going into an unspoilt place—a beach, mountain, or desert—we sense serenity, an order that transcends even the physical beauty. There are countless beings occupying many worlds and planes, most of them invisible to us. Existing in different frequencies of vibration, their home worlds may be passing through ours now, in the same space, yet none may ever interact with the others directly. We simply are not conscious of them, or they of us. All these planes and worlds are connected *in essence* with everything in nature, even though we may never perceive the bonds, especially with our physical senses. We nonetheless are affected by these interconnections, and we also affect them.

But with the course of time we have forgotten this great endowment by Mother Nature and have started taking these for granted. This has led to overcrowding, pollution, accumulation of toxic xenobiotics, oil-spill havocs that have shattered the pure and clean lap of Earth. In the words of Aldo Leopold, "We abuse land because we regard it as a commodity belonging to us. When we see land as a community to which we belong, we may begin to use it with love and respect." The environment is progressively changing as it is getting hotter, stormier, unequally crowded, more violent, and less biodiverse. New-age advancements in science and technology are responsible for the present day condition of the earth. In the words of Alan M. Eddison "Modern technology owes ecology an apology."

Spiritual ecology is part of an immense unfolding scheme. From the one divine homogeneity countless beings burst forth. Differentiation involves bipolarity, spirit and matter working together, blended as two aspects of one consciousness. W.Q. Judge gives a concise idea, in *Echoes of the Orient* referring to "'the essential Unity of all life and being'. Manifestation of life is differentiation of this unity, the purpose of differentiation is evolution, and the destiny of evolution is the return of all manifestation into its source and original unity".

Every entity is a spiritual ecosystem composed of smaller lives, just as our bodies are made up of countless lives-atoms, cells, organs-working in harmony and held in beautiful order and balance. They are worlds and universes real and substantial in their own right, and remain mostly undetected as such. Just as atoms and cells are part of us, we belong to the living earth, Gaia, which is incorporated in a living solar system. This cell is one tiny part of a galaxy belonging to super clusters of galaxies.

In a more inclusive context we define spiritual ecology as a complex and diverse arena of intellectual and practical activities at the interface of religions and spiritualities on the one hand, and on the other of ecologies, environments, and environmentalisms. The term spiritual ecology is preferred over religious approaches towards ecological or natural conservation for four reasons:

- It parallels the well-established primary components of contemporary ecological anthropology like cultural ecology, historical ecology, and political ecology.

- It is more inclusive than the term religion.
- Other terms, such as religion and ecology, religion and nature, and eco-theology are awkward or problematic as they tend to give a biased touch to this concept.
- The term spiritual ecology can be used to be provocative (Schneiders 1989).

Since the 1960s there has been growing interest in spiritual ecology, and this interest accelerated markedly in the 1990s with an increasing number of conferences and publications. Spiritual ecology focuses on the relationships between religions and environments from the local to the global levels to address environmental crises, problems, and issues. This reflects the growing realization of numerous and diverse individuals and organizations that greener environmental world views, attitudes, and values are the most fundamental prerequisite for resolving environmental concerns, and that this in turn requires a profound rethinking of the meaning of nature and of the place of humans in it. Thus spiritual ecology operates at two levels, theoretical and practical. The latter includes religious organizations as well as individual spirituality.

Practice of Spiritual Ecology

The followers of spiritual ecology are people who preach and teach the interdependence of every living thing. They have an ethic based on the sacredness of life. They see it as part of mankind's duty to help preserve a cosmic balance as an essential aspect of all existence provided by the divine power of the Great Spirit. They feel that every war, illness, or misfortune is due in some way to a loss of this balance, and try to restore it in part by their ceremonies and in their everyday life. In our day to day life the practice of spiritual ecology involves three important steps both at the individual as well as at institutional level. These are: Awareness, Gratitude and Spiritual activism.

The awareness that the birds, mountains , animals, insects, their business and play in the world, their unique relationships to each other, each took 14 billion years of evolution to shape the first step in the practice of spiritual ecology. It is an awareness of the larger, ancient rhythms around us, of the seasons and the long story which has shaped us, and a vivid experience of our connection to everything else.

The second step in practicing spiritual ecology is gratitude. Food, shelter and energy are our most basic needs. All three of these needs have evolved from the creativity of the planet, from the demands placed on us and the gifts given to us by the Earth and are met with abundance in our lives. An understanding of these gifts and a humble appraisal of our Mother Earth can go a long way in promoting the most basic ideas of Spiritual ecology. Spiritual activism is the third and final step in the practice of spiritual ecology. It means informing ourselves on the issues and taking action on the side of justice and compassion. On a simpler note it is the stimulation of human minds to arouse the beings to work for the conservation of the delicate ecological balance prevalent in our nature.

Attaining Spiritual Ecology

The world's wisdom, traditions and scriptures provide an important guide to discovering and earning our true place in the universe. They tell us about this endless divine ecology where we have experienced innumerable forms and states of consciousness. We are so close to realizing these truthful ways of living in harmony. The great teachings in all religions remind us that the Creator has made us as one whole: we are one being. There is no separation, except in our thinking. So, to journey and make these profound movements in perception-to open our perceptions to the truth-is to know that we are connected and able to communicate with, and be in harmony with, all things in the circle of life. This is really the basic message of the sacred spiritual ecology.

The Buddha's life, from birth to Nirvana, was spent close to nature. He was born under a tree, attained enlightenment under a *bodhi* (pipal) tree and throughout his 45 years of teaching, he resided and preached in the forest. And even when he was dying, he lay down under a tree. Buddhist philosophy and values not only provide a good model for individual development, but also serve as a great example for how it can be applied to and benefit community-based education and conservation tasks. Buddhism provides approaches to promote self-reflection and a healthy relationship between oneself and others, including nature.

Even the Christians all over the world, in rich and poor societies have taken notice of the spiritual importance of environmental problems. "Not only must we deal with the effect of industrial society on the quality of the air and water essential to life, but many religious people have recognized the "meaningless of their consumer society" and have a longing to reestablish a connection to the earth." St. Augustine, the Bishop of Hippo in the 5th century said that we feel the goodness of existence in ourselves, and we observe it in all living beings.

Sri Krishna in the Bhagavad Gita says, "I pervade the Universe. All objects in the Universe rest on me as pearls on the thread of a garland." He also says "Conserve ecology or perish". This message of the Bhagavad Gita is a dialogue between Sri Krishna and Arjuna and is clear and precise Life Science. It is narrated in the third chapter of this great work that a life without contribution toward the preservation of ecology is a life of sin and a life without specific purpose or use.

In the words of the holy Islamic book Quran, "Do you not observe that God sends down rain from the sky, so that in the morning the earth becomes green?" (Sura 22:63). The color green is the most blessed of all colors for Muslims and, together with a profound sense of the value of nature as God's perfect and most fruitful plan. It clearly provides a charter for a green movement that could become the greatest exertion yet known in Islamic history, a "green *jihad*" appropriate for addressing the global environmental crisis.

Another approach towards attaining spiritual ecology is visiting sacred places which range along several continua (natural to anthropogenic, biophysical to socio-cultural, prehistoric to recent, etc.). These places are special because of some extraordinary attributes that stimulate feelings of power, mystery, awe, transcendence, peace, and healing. The same sacred place may be recognized as special by individuals from very different cultures, religions, histories, and ecologies, although they may interpret it quite differently. Many sacred places, particularly in nature, may promote environmental and biodiversity conservation, since local communities recognize them as special and prescribe certain behaviors in relation to them.

It is in spiritual awareness that the role of religious institutes and organizations become evident. They can motivate the people to care for all the life forms present on earth. They can be instrumental in enlightening minds and broadening the vision of human minds towards the spiritual aspect of nature by regarding it equivalent to us in being the creation of the one God. In the end, it can be said that we must be collaborators with nature, sharing a great impersonal love and sympathy for all beings, for true ethics can be based only on spiritual oneness and universal brotherhood. Albert Einstein has some wonderful thoughts concerning the personal self and our liberation - "Our task must be to free ourselves from the prison by widening our circle of compassion to embrace all living creatures and the whole of nature and its beauty. Nobody is able to achieve this completely, but the striving for each achievement is in itself part of the liberation and the foundation for inner security."

By bringing out a more fully conscious realization of our divinity we will be able to become more receptive to the vast inner currents and meanings of life, and know our true place in life's spiritual ecology. The outcome of this is to live unselfishly for the benefit of all beings. To summarize, the challenge before us is to establish a planetary society that is peaceful, just, and sustainable. In the words of Guy Dauncey "Maybe it is a sign of planetary maturity that people learn to respect the atmosphere as a sacred part of nature, not as a commodity." This is the right time to act for the entire Kingdom of life on Earth, for a new and vibrant earth community driven by the three-fold urge: "to live, to live well, and to live better."

References

B.M. Eagle (2002), *Sacred Ecology and Native American Spirituality*, the Rocky Mountain Spiritual Emergence Network Times, Vol. 5, No. 2.

Lauren de Boer (2002), *Energy and the Practice of Spiritual Ecology*, Earth Light Magazine, Spring, Vol. 45.

Kaveh L. Afrasiabi (1995), Toward an Islamic Ecotheology, *Hamdard Islamicus*, XVIII, No. 140.

Abraham, Ralph (1990), "Orphism: The Ancient Roots of Green Buddhism." In *Dharma Gaia: A Harvest of Essays in Buddhism and Ecology*, ed. Allan Hunt Badiner, 39–49. Berkeley, Calif.: Parallax Press.

Cremo, Michael A., and Mukunda Goswami (1995), *Divine Nature: A Spiritual Perspective on the Environmental Crisis*. Los Angeles, Calif.: Bhaktivedanta Institute.

Schneiders, Sandra M. (1989), "Spirituality in the Academy," *Theological Studies* 59 (4): 676-697.

Nollan, Jim (1990), *Spiritual Ecology: A Guide to Reconnecting with Nature*, New York, NY: Bantam Books.

Sponsel, Leslie E. (2005), "Ecological Anthropology," in *Encyclopedia of Religion and Nature*, New York, NY: Continuum Press. 1: 544-548.

Gardner, Gary (2002), *Invoking the Spirit: Religion and Spirituality in the Quest for a Sustainable World*, Washington, D.C.: Worldwatch Institute, Paper 164.

Posey, Darrell, *et al.*, eds. (1999), *Cultural and Spiritual Values of Biodiversity*, London, UK: Intermediate Technology Publications/UNEP.

CHRISTIANITY AND THE SURVIVAL OF CREATION

Piku Dasgupta

"Creation is the world in which we daily encounter bread and beauty, majestic mountains and familiar neighborhoods, painful brokenness and solace for the spirit. Creation is where we stand as we hear the divine summons to care for our neighbor and for the earth. Creation means that all things in heaven and earth are related to the One who gives them their beings. Creation means that we are creatures too; the healing of the earth and the healings of human persons and human society must go hand in hand." But humans have undeniably and boldly placed themselves and all other life in ecological crisis. We have not only failed to preserve the diversity of species that inhabit the earth, but we have failed to recognize the value of diversity that is implied in biology as being inherently valuable within our species. We have created ghettos with our "isms" and placed the value of intolerance among religions, gender, races, etc. If we look to our religions to establish a system through which we define our moral praxis, our spiritual identities, we must all redefine our purpose as humans on earth.

We need to explore the relationship between religion and the environment, between Creator and Creation. This area of religious study is perhaps esoteric in nature, but is applicable to all people, even for those who are not religious. This seeks to serve as an introduction to the study of religion and its environmental implications. Ecology, eco-feminism and eco-womanism, despite their particulars of whom to blame for the destruction of the earth and which aspects should be focused on, they are essentially different knots on the same rope. For deep ecologists, it is the anthropocentric worldview that is foremost to blame for environmental destruction and eco-feminists on the other hand, argue that it is the andocentric worldview that deserves primary blame–that it is not just humans, but men and the masculinist worldview that must be dismantled from their privileged place. All views are valid among these, but it seems more important that we act for the earth and all her creatures rather than discuss semantics. Thus what is required is a more inclusive paradigm, one that not only attempts to liberate human vision from the small, ego-centred mind and break down walls of exclusion based on marginality, but one which attempts to de-ghettoize the new ecological movement.

We face an environmental crisis today of staggering proportions. All forms of life need and depend upon one another for their health and survival and the gradual destruction of one life form eventually results in a ripple-like effect, in the degradation of the whole eco-system that originally supported the life form now under siege. But we seem confused as to how to address the crisis in a manner that will engender long-term sustainable growth and development for human communities without sacrificing the vital needs of non human communities to survive and flourish. However we apparently lack the heartfelt commitment to sustainability. It required for ensuring the integrity of humankind and other kind in unity with one another. The vital questions before us are, why is the biosphere on which we daily rely upon for sustenance and refreshment, in such trouble? Is the cause of our collective inability to address adequately the earth crisis a cognitive failure to understand what it will take to build sustainable communities? Do we not know how to solve the problem? Do we care enough about interspecies integrity to feel motivated to address the predicament at hand? Probably the fundamental cause of our collective inability to confront the global environmental crisis is our deep seated unwillingness to change our habits and embrace greener lifestyles. But unless at the core we are fundamentally committed to sustainable living, no amount of eco-efficiency in business and industry will make for long lasting change.

The environmental crisis is a spiritual crisis because the continued degradation of the earth threatens the fundamental goods and values that bind human beings to one another and all other forms of life. At a very deep level we no longer feel our common kinship with other beings as the basis for earth-friendly action and commitment.

The earth crisis which is a spiritual one also means that the problem is explicitly a religious problem in the sense that the promulgation of particular theological teachings has led to the ravaging of earth communities. For example the idea in the Genesis Creation story that God, a heavenly being, far removed from our planet, created human beings as God's vice-regents to exercise "dominion" over earth. If God has given the earth to us as our private possession, then why not do with it what we want to? According to Lynn White, Western Christianity's attack on paganism effectively stripped the natural world of any spiritual meaning by replacing the belief that God is a disembodied spirit whose true residence is in Heaven, not on earth. The impact of Christianity's anti-pagan teachings has tended to empty the biosphere of any sense of God's presence in natural things. God is now pictured as a Sky-God with little if any connection to natural processes. These teachings inculcate in human beings an absence of family feeling for other biotic communities. In this sense therefore the ecological crisis is fundamentally a spiritual crisis because certain Christian teachings have blunted our ability to experience co-belonging with other life forms, rendering us unwilling to alter our self destructive course and plot a new path towards sustainable living. Therefore the question now is how might a reinterpretation of the Christian tradition help to resolve this crisis? Christianity should survive and renew itself so that it may become as largely and truly instructive as we need it to be.

Ecocide thus is a spiritual disease–a disease of the soul and requires the medicine of healthy, rather than toxic Christian values and ideas. This paradox is not surprising because it is often the case that the seeds for positive change remain deep within the very thing that is the source of the seemingly incorrigible impediments to change in the first place. The recovery of the ancient idea of the Spirit as God's presence within all things is directly applicable to one of the most intractable ecological problems in our times. In so far as the Earth Spirit lives with us in and through the created world, then God as spirit suffers loss and pain whenever the biotic order is despoiled through human arrogance.

In Western religious thought, Spirit is defined as a bodiless, non-material reality, over and against the physical world. Western thought generally operates according to a series of binary opposition that separates spirit from body, mind from matter and God from nature. Therefore Western thought has not only pitted the spiritual world and the physical order against one another but also subordinated the one to the other. The biocentric role of the spirit as the power of the life-giving breath within creation, including non human as well as human creation has been consistently downplayed. To reconceive the Spirit as the enfleshment of God's sustaining power in the biosphere is to emphasis the co inherence of the Spirit and the natural world. Spirit and earth internally condition and permeate one another; in so far as the spirit abides in and with all living things, spirit and earth are inseparable and yet at the same time indistinguishable.

Spirit and earth are internally indivisible because both modes of being are living realities with the common goal of sustaining other life forms. But spirit and earth also possess their own distinctive identities in so far as the spirit is the unseen power who vivifies and sustains all living things while the earth is the visible agent of the life that pulsates throughout creation. Under the control of this dialectic, the earth is the body of the spirit. The spirit inhabit the earth as its invisible and life giving breath *(ruah)* and the earth *(gaia)* is the outward manifestation, the body, as it were, of the Spirit's presence within the maintenance of all life forms. Green spirituality affirms the common interdependence of all persons with each other–indeed of all species with each other–as we all struggle to protect the integrity of the life web that holds together our planet home. In so far as the spirit breathes into and sustains life for all members of the web, green religion testifies to the bond of unity that unites all God's children together on a sacred Earth.

The spirit in our time is the "wounded spirit" or "cruciform spirit", who, like Christ, takes within itself the burden of human sin and the deep ecological damage this sin has wrought in the biosphere. As Christ's wounds become the Eucharistic blood that nourishes the believer, so also does the spirit's agony over damage to the earth become a source of hope for communities facing seemingly hopeless environmental destitution? One of the many ironies of Christian faith is the belief that out of death comes to life, from loss and suffering comes to the possibility of hope and renewal. This irony is symbolized in the creator's emptying of herself in creation so that all beings may enjoy fullness of life. Our rapacious habit daily would afresh the Earth Spirit that breathes life into all things; and daily the Earth Spirit intercede for us and protects us by allowing us to remain richly alive in spite of our behaviour to the contrary.

One of the most compelling Christian responses to the threat of ecocide lie in the recovery of the Holy Spirit as God's power of life giving breath *(ruah)* who indwells and sustains all life forms. An answer to the increasing environmental degradation in our times is not better technology–but a spirit-motivated conversion of our whole ways of life to sustainable living. Such a change of heart can occur through an encounter with Christian wisdom. Today we need a conversion of heart to an integrated planetary vision of a green earth where all persons live in harmony with their natural environments. May the Holy Spirit, come into our hearts and minds and persuade us to work towards a seamless social environmental ethic of justice towards all God's creatures. Christianity requires a new, all inclusive Christic paradigm, one which has at its essence, the earth not as a "backdrop" of salvation. At the very beginning of this paper it was showed that Christian goals require that man not be of the earth, but to rise from it, to escape from its imperfections. But rather than viewing God as gendered, dominating and watching us from a distance we might regard God not as being one with whom we should have a relationship, but we might imagine God as the relationship. If we envision God as life and as the thread which makes all living things inter dependent and inter related, we can now define what sin might be, what can be the greatest fallacy a human could commit against God: interrupting the web! The hope for a renewed earth is best founded on belief in the Spirit as the divine force within the cosmos that continually indwells and works to sustain all forms of life. In this way ancient earth wisdom within Christianity can play a role in making our respective bioregions vital places in which we live and work.

Notes and References

1. Leslie Roberts et al., World Resources 1998-99: A Guide to the Global Environment (New York: Oxford University press, 1998) and Jeremy Rifkin, Biosphere Politics: A cultural odyssey from the Middle Ages to the New Age (San Francisco: HarperSanFRancisco, 1991).
2. Lynn White, Jr., "The Historic Roots of our Ecological Crisis," Science 155(1967): 1203-7.
3. The Ecological Community: Environmental Challenges for Philosophy, Politics, and Morality, ed. Roger S. Gottlieb (New York : Routledge, 1997), 292-310.
4. Jurgen Moltmann: The Crucified God, trans. R.A. Wilson and John Bowden (New York: Harper & Row, 1974) 244.

DEATH: PHILOSOPHY VERSUS FACTS

Aarti Kanojiya

Death and life after death have always fascinated human. The term 'death' is ambiguous. The ending of life is one thing, and the condition of having life over is another. 'Death' can refer to either. Death is a process of ending of life, the loss of very trace of life. It is a state of process of extinction denouement, i.e., final completion. But the actual definition of death is still not clear. The mirror image of death is birth (more precisely becoming alive), and the two affect us in very different ways: birth makes life possible; it starts a good thing going. Death makes life impossible; it brings a good thing to a close. As Frances Kamm (1998) emphasizes, *'we do not want our lives to be all over with'*.[1] Death is termed as end of life but still it is not clear what life is. Thus an attempt has been made in this paper to deal with philosophy and real facts about death.

Religious Beliefs about Death

Death is described as a process of the soul rejecting its old body to take up a new one. Death is a process by which a spirit passes from one body to another gaining experience in life. According to Christian religious traditions life does not permanently end, when bodies break down. Life may continue in two ways - God may resurrect our bodies and restore our mental life or life may continue uninterrupted assuming we are souls who survive the demise of the body and after death soul departure from body.[2] Person may not exist as person after death or end of life. They survive as carcasses or corpse (thus die). Persons may exist after their lives end; we mean to discover if they may survive *as persons* after their lives end. Perhaps, insofar as we are animals (as animalists, such as Snowdon 1990 and Olson 1997 suggest), we may persist for a brief time after we are no longer alive, given that animals persist for some time as corpses or carcasses, according to some theorists.[3] But even if we may exist after death *as animal corpses*, it does not follow that we may exist after death *as persons*. Assuming that personhood entails the possession of psychological attributes, and possessing these requires being alive, we cannot continue to exist as persons when our lives end.

Thus, after death personhood is lost which entails the possession of psychological attributes. The ideas of death have undergone various alterations in accordance with man's evolution, but the deep mystery of life after death was not explained until after the advent of Christ, who through his death upon the Cross brought to man the hope of salvation. The Bible says, (St. John, 5:24) "Verily, verily, I say unto you, he that hearth my words and believeth on him that sent me hath everlasting life, and shall not come into condemnation, but is passed from death unto life." The Apostle Paul states in the Bible (II Timothy 1: 9-10), "God hath saved us and called us, not according to our works, but according to his own purpose and grace, which was given us in Christ Jesus before the world began. But is now made manifest by the appearing of our Savior Jesus Christ, who hath abolished death, and hath brought life and immortality to light through the gospel." With Christ's supreme sacrifice the world has grasped the opportunity to bring the hope of immortality.

Philosophy behind Death

Philosophers say that the soul is formed by a combination of the five elements. Combination of atoms and molecules can never generate *intelligence of consciousness.* Motion cannot produce sensation, ideas and thoughts. These are attributes of a universal soul. Only a combination of body and soul can generate this. The physical body is constantly changing but the soul is eternal; it is the basis and source of matter and energy. The soul can never be demonstrated, but its existence can be inferred by certain empirical facts. Death is distribution of our entity. Hence further clarifying what it is for a person to die entails clarifying what is essential to a person's identity,[4] how identity is instantiated in particular structures, and what is involved in the breakdown of those structures It is beyond the realm of physical and material science. Man is a soul bearing a physical body. Soul is an immortal part of life. At death, the desire and mental bodies leave physical body and take with them one permanent atom that is deposited in the left ventricle of brain. It is like a negative film of a camera and is impressed with all experiences of the life that just has ended. At death the force of this atom leaves the body and all these impressions are transferred from the vital (etheric) body (which is the storehouse of these experiences) into the desire body, which then forms the basis of the man or woman's life in purgatory and the first heaven. This transfer is done by the spirit during the first three and one- half days after the rupture of the connection between the seed atom and the heart, ordinarily known as *Death*.

We may thus see that death is not complete until this transfer has been accomplished. Sensation is still present, and the spirit suffers through inharmonious surroundings. It can feel somewhat during postmortem examination or embalming. When the body is mutilated or cremated before the silver cord is severed, the spirit suffers pain. The doctors and undertakers, believing the person "dead," usually do not handle the body with the same care that they would if they knew the real facts. For example, Feinberg (1984), following Levenbook (1984),[5] defines death as "the first moment of the subject's nonexistence." Perhaps this definition is motivated by the awkwardness of attaching 'death' to a moment in the dying process when a spark of life persists. However, it is at least as awkward to attach 'death' to a moment after the dying process is over to suggest that the ending of life occurs while we are in a state of death.[6] As the interest and belief in a life after death becomes more universal, the necessity for a scientific method for the care of those who are passing into the higher life will be impressed upon the people, and we shall then have nurses, doctors and ministers who are versed in the science of death as well as science of birth. The spirit will then be surrounded not only with love, but with peace and quiet at the time of passing. It will also have a deeper and clearer record with which to begin its life work in its new state.

Scientific Aspect of Death

Since World War I mankind has become greatly interested in life after death. But all what is mentioned in philosophy is contrary to our experience. Our experience says that soul never wants to leave a body rather it clings to its own body. Sir Oliver Lodge says the *ether* is in fact what the religionists call the "spirit" world. We have had the experimental proof that the so-called spirit world exists ever since 1874 when Sir William Crookes published the results of his experiments in the Quarterly Journal of Science. This has since been repeated by international teams of scientists and by lay people to the point of boredom. What we have been waiting for and what has been missing is a mathematical theory to back up such revolutionary[7] scientific experiments.[8] There is a very good reason why the religionists and their powerful materialistic allies are fighting like tigers to destroy the secular scientific case for survival after death. This is because whenever we speak to people who once lived on Earth and are now living in the normally invisible part of the universe, the message is always the same via mediums, it never varies: "There is no place here just reserved for Christians or anybody else, we all survive the death of our physical bodies."[9] As we are making a study of forces that are out of range of our five physical senses, we have to work with a machine that will bring into focus what we cannot normally see and sense.

There is nothing supernatural or paranormal about a microscope. It is a medium, a go-between, proving that just because we cannot see or sense something it does not mean that it does not exist. It is the same with radio and television sets. It is no good climbing on our roofs and trying to see the invisible forces hitting our aerials. We have to work with a machine, a medium that can do the job.

In order to make contact with people in the invisible part of the universe we have to work with a very sophisticated machine indeed. Something far more complicated and advanced than anything invented by human beings-the human mind and brain. This is why we have to work with a materialization medium in order to prove that we all possess a soul that separates from the dead physical body. For over 100 years we have had the experimental scientific proof that we all survive the death of our physical bodies, and that the living mind separates from the dead brain. Sir William Crookes, later to be awarded the Order of Merit and to hold the office of President of The Royal Society, carried out repeatable experiments under laboratory conditions. His results were published in The Quarterly Journal of Science in 1874. These experiments proving survival were subsequently repeated by international teams of scientists. One of the main reasons why this revolutionary discovery in physics failed to have the impact that it should have done is that these experiments lacked the mathematical theory to back up what was being observed. Thanks to recent exciting discoveries in subatomic physics, the study of the invisible part of the universe, and the published work of UK Scientist, Ronald Pearson, we now have the missing mathematical theory that matches the pioneering experiments.

The highly developed Medical Science, the innumerable Medical establishments and the various methods of treatment prove that nobody welcomes death. Death does not take place at the will of man, rather it is an involuntary act, and something imposed upon man from which he cannot escape. People are willing to pay any price if they can overcome death. If an old man dies and carries with him all the experiences he has gained and takes up a new body, the new infant should behave like an experienced man. But this does not happen! Derek Parfit strongly proposed their views such as memories and character gradually change over time.[10] Death too come in degrees and become complete when our psychological profiles are greatly altered or destroyed. If we have continual identity then death is all or nothing.

Thus death is not soul casting off its body to take up a new one; it is actually ruling over the body. For whatever is the reason, we humans die. By scientific investigations a pretty good idea of why death occurs is gathered. Cells die without oxygen. Starvation and stomach or intestinal problems are the ways one can die in which lack of oxygen may not be the direct cause of death. In all other cases be it a gunshot wound or cancer, it comes to lack of oxygen. Some people die of fatal illness or coma, cells without oxygen, starvation, intestinal problems are causes of death. Death begins when heart stops beating. Cells devoid of oxygen like brain and skin cells die. Thus death is not an event; it is a process rather than an event.

For example, if an individual is frozen and later revived, as is sometimes done to simple organisms, it is tempting to say that he ceases to be alive while frozen but is in a state of suspended animation. Or imagine a futuristic device that reduces to disconnected atoms which it stores and later reassembles just as they were before. Many of us will say that I would survive – my life would continue – after the reassembly, but it is quite clear that I would not live during intervals when my atoms are stacked in storage. In these cases, our linguistic intuitions give no definitive verdict concerning the applicability of 'death'. On the one hand, it seems appropriate to say that I die when my body is completely frozen or my atoms are disconnected, since the term 'death' seems applicable when a creature's life ceases.

On the other hand it seems correct to deny that I die, since my life is eventually restored, and 'death' seems applicable only when a creature. Second, it is important to distinguish between the concept of death and a criterion for death. The concept of death says what death is. One such concept is that of the cessation of personal survival. A criterion for death, by contrast, lays out a condition that is sufficient for death and by which an individual's death may be determined. The traditional criterion says that you will be dead when your heart and lungs cease to function (not that death *is* cessation of respiration and cardiac functioning). A more recent criterion is brain death – meaning the death of the entire brain – since the brain is the seat of our psychological features.

The brain death criterion is more accurate since, with modern technology, respiration and blood circulation can be maintained artificially even when the brain is dead. As things stand, authorities in the legal and medical context frequently rely on the brain death criterion (President's Commission, 1981). For example, tissues are not to be harvested from organ donors unless the entire brain is dead. But there is good reason to consider a person dead even if certain parts of the brain are still alive. Psychological attributes are most closely associated with the higher brain (the cerebral cortex). Unsurprisingly, then, there is increasing support for a *higher* brain criterion for death, according to which death occurs when the higher brain is no longer alive. But while higher brain death is a sufficient condition for a person's death, it is not a necessary condition. Conceivably, the higher brain might live through damage or alterations that destroy our psychological profile; if so, we might die as persons while the higher brain remains alive.

Conclusion

In the end, we may conclude that death is certain for the born. Before birth and after death beings are not to manifest human senses. They have to manifest only between birth and death. Are all deaths misfortunes? Perhaps, but there is a strong case to the contrary. To support the conclusion that death is not always a misfortune, we might adopt the desire satisfaction theory, according to which we are harmed by what thwarts our salient desires. Perhaps it is not bad to die at an advanced enough age, for people who live long enough may be ground down by life until they give up many of their goals. Also, they will have attained many of their aspirations. If already satisfied, or given up, a desire cannot be thwarted, even by death, so as we lose our motivation for living, death ceases to be objectionable to us. Perhaps death is bad for us only if it is premature in the sense that it comes when we still have interests such as salient desires that propel us forward in life, and only if meeting these interests is a real prospect. We are left to wonder whether death would ever cease to be objectionable were we *not* ravaged by bad health and other hardships. Bernard Williams[11] argues that it would be bad to live forever, even under the best of circumstances. His view is based on an assumption about the relationship between our identities and the desires that motivate us to live.

Consider a woman who wants to die. She might still take the view that if she is to live on, then she should be well fed and clothed. She wants food and clothing on condition she remain alive. In this sense her desires are *conditional*, and do not give her reason to live. Contrast a father who is committed to rearing a beloved daughter: he desires unconditionally that the child do well, and his desire gives him reason to live, because he can rear his child only if he survives. In this sense, his desire is categorical, or unconditional. Williams thinks that categorical desires are essential to identity, and give meaning to life. Through categorical desires, we are attached to projects or relationships that are definitive of the self; faced with their destruction, we would feel our lives are meaningless, and that in an important sense we cannot survive as the persons we once were.

The bearing on death, according to Williams, is, first, that people have good reason to condemn a death that is premature in the sense that it thwarts their categorical desires.[12] Second, mortality is good, since people who live long enough eventually will lose the categorical desires with which they identify. Life will lose its novelty, and oppressive boredom will set in. To avoid ennui, superseniors would have to replace their fundamental desires, again and again. But this is to abandon their identities; it is tantamount to death.

As Williams says, lives of unimaginative routine will eventually grow stale if extended long enough. Of course, this is not supposed to comfort ordinary mortals, most of whom will die long before routine undermines the joy in living. However, as several theorists, including Nagel[13] Glover[14] and Fischer[15] have suggested, it is not obvious that life must become dull. Williams may have overlooked how rich and complex life can be, especially for superseniors who pursue multiple open-ended projects in the company of other superseniors. His response to this kind of criticism is that even rich and open-ended projects eventually will become routine (say after a few billion years), so our pursuits must be replaced periodically if we are to remain interested in life. But to phase in wholly new projects is to lose our identity.

Notes

1. Kamm, F.M., 1988. "Why is death bad and worse than pre-natal Non-existence?" *Pacific Philosophical Quartely* 69; 161-164.
2. Mathew M.G., Yoga of Knowledge. Sankhya Philosophy. Geetha and Bible: 53-69.
3. <http://plato.stanford.edu/entries/death>
4. Green, M. and Winkler, D., 1980. "Brain Death and Personal Identity." Philosophy and Public Affairs 9,105-133.
5. Levenbook, Barbara, 1984. Harming Someone after his Death. *Ethics* 94, 3: 407-419.
6. Michael Roll, "A Rational Scientific Explanation for So-Called Psychic Phenomena", October 2004.
7. Ronald Pearson, "Survival Physics", October 2005.
8. Feldman, F.,. 1992 *Confrontations with the Paper.*New York ; Oxford University Press.
9. BBC Points of View notice board-Science and the Séance broadcast on BBC2 on August 31, 2005.
10. Parfit , D., 1984. *Reasons and Persons.* Oxford; Carendon Press.
11. Sivananda. S. S. What Is Death? Philosophy of Flesh.
12. <www.cfpf.org.uk and www.victorzammit.com>
13. Fischer, 1994, "Why Immortality is not so Bad," *International Journal of Philosophical studies* 2: 257-70.
14. Glover, J., 1977. *Causing Death and Saving Lives.* Harmoindsworth: Penguim Books.
15. Nagel, T., 1979. "Death" in Nagel, T., *Mortal Questions.* Cambridge: Cambridge University Press.

DEATH–THE DESTINY OF LIFE

Dipa Bardhan

> *"Death is a downfall, and only by faith can this downfall be interpreted as a falling into the hands of the living God."*
>
> —Karl Rahner

It has been claimed that one can never look directly at the Sun or at one's own death. And yet, throughout the history of mankind, both have been the enduring themes of myth and religion, science and magic, curiosity and fear. From our late twentieth century vantage point we find that as the Sun is understood as being the source of life in the natural order, so death is becoming recognized as the central dynamism underlying the life, vitality, and structure of the social order. Death is the muse of our religions, philosophies, political ideologies, arts and medical technologies. As Richard Huntington and Peter Metcalf observed in *Celebrations of Death* (1979:2), "*life becomes transparent against the background of death*".

Death is a catalyst that, when put into contact with any cultural order, precipitates out the central beliefs and concerns of a people. Apparently from the very beginning, death has presented a problem to humankind and human beings have had the equipment with which they foresee death and do try to understand it. There have usually been two problems in understanding death: on the one hand, the belief that life is rendered meaningless if it is cut off in death and, on the other hand, the belief that human beings, unlike other animals, are too valuable to suffer the fate that awaits all other living forms. There are three responses which mortals can make in regard to death: to deny it, to accept it as an unpleasant but inevitable fact, or to affirm it not only as inevitable but also as a valid and joyous part of the natural process of which birth, living and death are equally important. The response varies from person to person.

Theological Perspective of Death

Life and death are concerned with the eternal process of becoming, the process of growth, transition, regression, transformation: a process where the unexpected is forever breaking through the pattern of the seemingly established, and at every turn of the river of life new vistas may be opened enabling the soul to glimpse the country where ultimate mysteries, unknown and unknowable abide.

Death and Hinduism

"Our soul never dies; only the physical body dies. We neither fear death nor look forward to it, but revere it as a most exalted experience. Life, death and the after life are part of our path to perfect oneness with God." Thus, Hindus know death to be the soul's detaching itself from the physical body and continuing on in the subtle body (sukshma sharira) with the same desires, aspirations and occupations as when it lived in a physical body. Now the person exists in the in-between world, the subtle plane, or *Antarloka,* with loved ones who have previously died, and is visited by earthly associates during their sleep. They do not fear death, for they know it to be one of the most glorious and exalted experiences, rich in spiritual potential. Other terms for death include *panchatvam* (death as dissolution of the five elements), *mrityu* (natural death), *prayopavesha* (self-willed death by fasting) and *marana* (unnatural death, e.g., by murder).

For Hindus, death is nobly referred to as *'mahaaprasthaana'*, the great journey. "When the lessons of life have been learned and *karmas* reach a point of intensity, the soul leaves the physical body, which then returns its elements to the earth. Awareness, will, memory and intelligence continue to exist in the soul body. Death is a most natural experience, not to be feared. It is a quick transition from the physical world to the astral plane..." According to Sanskrit literature, death is a rich concept described as *mahaprasthana*, "great departure;" samadhimarana, dying consciously while in the state of meditation; *mahasamadhi*, "great merger, or absorption," naming the departure of an enlightened soul.

Another major aspect of Hindu dharma that is common to practically all Hindus is that of the *purusharthas*, the "four objectives of life". They are *kâma, artha, dharma* and *moksha*. It is said that all beings seek *kâma* (pleasure, physical or emotional) and *artha* (material wealth). Of course, the only goal that is truly ultimate, whose attainment results in ultimate happiness, is *moksha* (salvation), also known as *Mukti* (spiritual liberation), *Samadhi, Nirvana*, or escape from *Samsara* (the cycle of birth and death). Hinduism believes in the rebirth and reincarnation of souls. Death is therefore not a great calamity, not an end of all, but a natural process in the existence of soul as a separate entity, by which it reassembles its resources, adjusts its course and returns again to the earth to continue its journey. In Hinduism death is a temporary cessation of physical activity, a necessary means of recycling the resources and energy and an opportunity for the soul to review its programs and policies.

Death and Christianity

"Dust thou art, and unto dust shalt thou return" (Bible-Genesis 3:19). For Christians whose lives are guided by the Bible, the reality of death is acknowledged as part of the current human condition, affected by sin. There is "a time to be born and a time to die." Although eternal life is a gift that is granted to all who accept salvation through Jesus Christ, faithful Christians await the second coming of Jesus for complete realization of their immortality. While waiting for Jesus to come again, Christians may be called upon to care for the dying and to face personally their own death. When Jesus died on the cross he was giving up his life for other people's sins and demonstrating the love of God. When Jesus rose again he proved that it was possible to overcome death and live with God for all eternity.

Jesus said, '*I am the resurrection and the life. He who believes in me will live, even though he dies.*' Many Christians believe that one day Jesus will return to earth. This second coming of Jesus is called the *parousia*. At the *parousia* God will judge everyone and decide whether they will be given eternal life with God or whether they are to be punished in hell. Many Christians have different ideas about heaven and hell. Some follow a very traditional belief that heaven is a wonderful garden paradise where they will live for ever with God. These people often believe that hell is a real place where people will suffer eternal torment and punishment. Other people believe that heaven and hell are really states. Heaven is the condition of being permanently united with God and hell a condition where you are permanently unhappy and unable to be with God.

Death and Islam

Allah said in the Qur'an, *"But those who believe and do good deeds. We will admit them to gardens (Paradise) in which rivers flow, lasting in them forever..."* (Qur'an, 4:57). People of Islam believe in the oneness of God and do good works, God has prepared an everlasting Paradise. Anyone who follows and believes this may enter Paradise. Muslims believe that this world is their one and only chance to earn this gift. Life involves preparing the soul to become worthy to pass through death, which is a transition, and progress to eternity in Paradise. Allah will judge each person according to his or her beliefs and acts during this life. Each person will be judged according to his/her individual situation, beliefs, and acts towards others. Those who live according to the truth and Qur'an to the best of their ability will achieve the reward of Paradise. However, Allah warns: those who were shown the truth of Islam and rejected it, will be shown no mercy and shall face hellfire. Just as Paradise is a place of delight, Hell is a place of eternal torment. In Islam, a person's fate is completely contingent on what one does in this life. Visiting the graves is recommended for the living to remember death and the Day of Judgment.

Death and Buddhism

"Even death is not to be feared by one who has lived wisely," says Buddha. From its inception, Buddhism has stressed the importance of death, since awareness of death is what prompted the Buddha to perceive the ultimate futility of worldly concerns and pleasures. Realizing that death is inevitable for a person who is caught up in worldly pleasures and attitudes, he resolved to renounce the world and devote himself to finding a solution to this most basic of existential dilemmas. A Buddhist looks at death as a breaking apart of the material of which we are composed. However, Buddhism does not look at death as a continuation of the soul but as an awakening. Dying and being reborn has been compared by some Buddhist as a candle flame. When the flame of one lit candle is touched to the wick of an unlighted candle, the light passes from one candle to another. The actual flame of the first candle does not pass over but is responsible for lighting the second candle. Buddhism asserts that all being live beyond the various fluctuations of this life. Death is merely a passage to rebirth in another realm such as the human world, a pure land or the flowering of the ultimate nature of the mind.

Death and Zoroastrianism

"Death is more universal than life; everyone dies but not everyone lives" says A. Sachs. The Parsees (Zoroastrians) do not cremate, bury or submerge their dead in water because they consider the dead to be impure, and Zoroastrian faith does not permit them to defile any of the elements with them. This desert ritual, which originated along with their faith in Persia more than 3,000 years ago and regards death not as the work of God but of the devil, dictates that the dead be left to vultures on hilltops.

It is common for Parsees to travel long distances to bring their dead to the Mumbai towers because prayers for the dead can only be said for those who have passed through its gates. Dead Parsees are carried on a bier to a ceremonial gate near the five Towers of Silence, where relatives hand them to pallbearers, the only people allowed inside. The black stone towers, about 36 metres high, are like three-tiered, open-air arenas where the men are placed in the outer circle, women in the middle and children in the innermost for the vultures to feed on. But with an average of three Parsees dying every day, the handful of vultures at the towers are overfed. Experts say about 100-120 birds would be needed to deal with the daily intake of bodies. The Zoroastrians, or Parsees, have installed solar reflectors in their Towers of Silence in Mumbai to help dispose of their dead after a decline in the number of vultures that scavenge their corpses in keeping with tradition.

Death and Judaism

"The question is not whether we will die, but how we will live" says Joan Borysenko. Judaism has stressed the natural fact of death and its role in giving life meaning. The fear of death, concern about the fate of our own soul and those of our loved ones, ethical concerns that some people die unfairly, all these and many other issues are discussed in Jewish literature. Since God is seen as ultimately just, the seeming injustice on Earth has propelled many traditional Jewish thinkers into seeing the after life as a way to reflect the ultimate justice of human existence.

Traditional thinkers considered how individuals would be rewarded or punished after their deaths. There are a few rare descriptions of life after death. Traditionalists gave the name *Gehenna* to the place where souls were punished. Many Jewish thinkers noted that since, essentially, God is filled with mercy and love, punishment is not to be considered to be eternal. There are, similarly, many varying conceptions of paradise, such as that paradise is the place where we finally understand the true concept of God. It is also possible that there is no separate Heaven and Hell, only lesser or greater distance from God after death. Judaism does not believe people who are Gentiles will automatically go to Hell or that Jews will automatically go to Heaven on their basis of their belonging to the faith. Rather, individual ethical behavior is what is most important.

Biological Perspective of Death

One could raise an almost inexhaustible number of questions; of which a few of them should be noted. At what point in the life cycle (or death cycle) would one wish to be stabilized? Considering the problem of over population it would seem necessary-if human engineering did not alter the reproductive cycle-to fix human life before puberty or after menopause. Can one imagine a valuable, or even interesting, existence if growth for everyone were halted at either of these two developmental levels? Who could really wish for the monotony that would be a necessary consequence of our being rendered immortal in this curiously naïve sense? Hans Jonas has pointed out (in *The Phenomenon of Life;* Dell, 1968) that the very essence of what we mean by life is mortality: "Mortality is the very condition of separate self-hood which in the instinct of self-preservation shows itself so highly prized throughout the organic world." How can life be envisioned as lacking the complete process of birth, development and death? Whatever else we mean life, we mean change; without it, what remained would be a mere semblance. Ethically, we hope that lives will change for the better. But whether or not this is the case, the capacity to choose, to react to new conditions is the process of living that makes possible the worth of living. Death and life are inexorably bound up with each other. If we deny one, we deny the other and will inevitably live poorly. According to Frances Wickes, 'biologically, death begins in embryo'. There is a constant sloughing off and generation of body tissues, a process that goes on throughout what we call life and continues after we die. Spiritually a similar process seems to be natural. Some of these deaths and births are more cataclysmic than others, but whether dramatic or gradual, the process of spiritual living is essentially the same as that of bodily living.

Necessity of Death

"Biological death is pivotal in the transformation of nature, and in growth of the personality", says Teilhard de Chardin. Death is necessary in this narrow world of ours. Despite the appearance that it is a harbinger of misfortune and grief, it is needed for human happiness. Our globe cannot conceivably support the offspring of mankind if the harvester Death was not ever at work in the vast field of human life, mowing down the rotating crops as they mature. There are indeed limits to the numbers which this earth is capable of holding. As long as we live in the limited space, there is no withdrawal from the busy world of earthly affairs. The world is too narrow to allow for the segregation of different generations. Even granting a body continual health, could everlasting happiness be captured in a life within the bonds of earthly space? As generation after generation rises around a person, will they not wish to withdraw from a society of younger minds who view the uses of the world from a different perspective? Hope the answers to these questions will certainly be found fully.

The Physiology behind Death

It is of importance to know what is involved in the death of the body. Ancient writings describe it as a process by which the spirit is released. The general teaching is that the conjunction of the body with the spirit depends on the motions of the heart and the lungs of the body being conjoined with the corresponding pulsations and animation of the spirit. For the spirit also has a body with a pulse and respiration. Death occurs when, from any kind of disease or accident, the body comes into such a state as to be unable to act in unison with its spirit and carry out its behests. What is called Death occurs when the vital motions of the lungs and the heart cease and the correspondence with the activities of the spirit's heart and the spirit's lungs is broken.

Death occurs from below-the forms or connections which first dissolve are the most external-those of the red blood. And life in the body depends on the circulation which supplies the tissues with vital heat. Physiology takes account of the fact that in organic bodies' heat is being maintained in calculable degrees of temperature and amounts of energy, through the consumption of oxygen by the tissues. When the heart stops beating and the body grow cold - which differs as to time according to the fatal disease -the bond between the spirit and the material body is dissolved.

Death occurs when the two vital motions, the respiration of the lungs and the beating of the heart, cease, and the body, deprived of the life of the spirit, grows cold and begins to decay. But until the heart's motion is entirely stopped the spirit continues in the body for a short time. And even after the body is apparently cold, life may with some persist as conscious thinking. The spirit can of course not have any sensation of its natural environment - since respiration has stopped-nor can it move even a particle of the gross matter of the body. The spirit, though definitely severed from the body, may still abide in it, which eventually "recedes" as a cutis-like covering.

Three phenomena may describe death. First, molecular death-in-life. Starvation carries the threat of death daily. The mammalian metabolic response to starvation is tripartite: glycogenolysis in hours, gluconeogenesis in days, and ketogenesis in months. All of these mechanisms of internal fuel regeneration are versions of catabolic molecular destruction. However, gluconeogenesis is distinctive in that it links destruction of muscle and adipose tissue to the creation of new glucose, the ubiquitous bioenergetic fuel. Dying and living, as destruction and creation, are here juxtaposed. Starvation is also a good spiritual metaphor.

Second, cellular death-in-life. The pathologist Andrew Wyllie first described the process of programmed cell death, or apoptosis, now recognised as central to embryogenesis, neurogenesis, immunoregulation, inflammation and tumorigenesis. In apoptosis, organized cell deletion allows new growth, development, and indeed survival.

Third, tissue death-in-life. The biological process of wound healing consists of three stages: wound stabilisation by "organised" blood clot, then tissue demolition, and finally tissue regeneration and in growth. Tissue macrophages are dominant in remodeling wounded tissue at each stage, being both destroyers and regenerators, agents of tissue death and life, as healing proceeds. The necessity of death within the evolutionary context is a theme of Arthur Peacocke. As he rightly says, "Biological death of the individual is the prerequisite of the creativity of the biological order." and "Biological death can no longer be regarded as in any way the consequence of anything human beings might have been supposed to have done in the past, for evolutionary history shows it to be the very means whereby they appear, and so, for the theist, are created by God". These perspectives on biological death are entirely positive and congruent.

Conclusion

"Death is the opening of a more subtle life. In the flower, it sets free the perfume; in the chrysalis, the butterfly; in man, the soul" says Juliette Adam. Nikos Kazantzakis in The Saviors of God (Simon & Schuster, 1960) states our ethical responsibility better than most when he describes the individual as seed. The seed has continuity with the past, but has one unique responsibility: to blossom as only that seed can and thus to "save God." But where we err is in the assumption that the fact on which meaning is based is eternal. We do not live forever, nor does any part of us. Yet our acts may, or may not, have eternal significance. It is essential that each of us be aware that we make-or rather, contribute to-what is eternal. We should not fear death, for then we cease to live fully now. Our desire not to die-to live forever or live again-must be put away lest we not live at all. Our need to be there in the future, to be "rewarded," vitiates our acts and turns them into ego trips instead of experiences of loving and living.

Moreover, as we know, what we are biologically was a product of a long existence before we were born, and in the process will nourish and develop long after we die. But our influence on our world also goes on producing changes long after we are forgotten. As we live, we nourish or poison the lives around us, although to most of us it is not clear how or to what extent or in what ways. But each of us knows in his or her heart that this is so. In our most conscious moments we feel our responsibility to be aware of the meaning we have for others.

The essential philosophical problem concerns what we mean by meaning. Or, to put it another way, religious significance must be described in terms that do justice to our human response to the distinction between what is essential and what is ephemeral. As Whitehead puts it in *Science and Philosophy,* "The world which emphasizes Persistence is the World of Value. Value is in its nature timeless and immortal. . . . The value inherent in the Universe has an essential independence of any moment of time; and yet it loses its meaning apart from its necessary reference to the World of passing fact. Value refers to Fact and Fact refers to Value." Alan Harrington in *The Immortalist* (Avon, 1970), maintains that "life could be made over to eliminate death altogether through human engineering" (genetic and otherwise). Going even further, he says that "the primary source of our fears, and all evil and meanness afflicting the human spirit . . . was death all the time, and nothing else." "Science says: "We must live," and seeks the means of prolonging, increasing, facilitating and amplifying life, of making it tolerable and acceptable, wisdom says: "We must die," and seeks how to make us die well"-Miguel de Unamuno y Jugo.

References

<www.findyourfate.com/ deathmeter/ religion-death>

<www.experiencefestival.com/ a/ Death/ id/ 59246>

<http: / / www.deathonline.net/ index.cfm>

Peacocke Arthur (1990), Theology for a Scientific Age pp. 62-63, SCM Press.

Rahner Karl, (1986). Dialogue pp. 246-247, Crossroad, New York.

Wyllie A.H. (1987), Cell Death (is) Cytology and Cell Physiology (4th ed) pp. 755-785.

G. H. Bourne (ed.), Academic Press, New York.

Erickson, Millard (1985), *Christian Theology*. Grand Rapids, MI.: Baker Book House.

Huntington Richard and Metcalf Peter, (1979) *Celebrations of Death* (2)

The Inner World of Childhood (1968). New American Library.

Holy Bible.

Holy Qur'an.

A THEOLOGICAL AND BIOLOGICAL DEFINITION OF DEATH

T. Kamala

Biology deals with the study of living organism in their environment. Earth is the only *bioplanet* where living fauna, flora, Human beings exist and *interrelate* to one another and also *interact* with their immediate surroundings. Thus, we understand both living and non-living are *inseparable* and *co-ordinately co-exist* in the universe.

Each organism is formed of many tissues, which are intern formed of (millions) the basic units called as cells. Each cell is potent to carry on a number of metabolic reactions (reduplication, respiration, synthesis of macromolecules etc.,) for life. Pathogens like bacteria and virus act on host cells and *damage their integrated organized system at cellular level*. Both pathogens and any extreme conditions like low and high temperatures, pH, Salt concentration or lack of essential nutrients may prevent reduplication or may even result in *disintegration* which could ultimately result in the *death of the organism*. Biologically death is defined as "the *absence of potential life*", more specifically death means "the lack of ability to synthesize new molecules at cellular level results in the death of the tissues and ultimately that leads to the death of the organism. Thus *death commences from the basic unit cell to the organ and then to the whole organism*".

Though one organism dies at maturity *it leaves its genetic material in its off springs*, through reproduction. The young ones *inherit* all the *information* it needs for its own structure and function from its parents. *Information is in the form of genomes*. The genome is the total set of genetic material carried by every cell of an organism. All cells are derived from pre-existing cells, eg: we human come from the *sperm* and *egg cells* of our parents and all the cells in our bodies are the products of repeated divisions of a *fertilized egg* and the cells derived from it. Thus *cells form the link from one generation to the next*. Contemporary cells and the organisms that are composed of them are parts of a *continuous living chain going back to the first simple cells on earth* (3½ billions years ago).

Thus we understand even after death, the *organism survives* through its *reproductive cells*. Both plants and animals have the ability to *perpetuate* themselves *indefinitely*. Most of the plants have *perennating* vegetative structures like, bulbils, gemmae, adventitious buds, protonema, etc. Though the entire plant body dies during unfavorable season, with the onset of favorable conditions, somatic cells of the plants by their *perennating* structures develop into a new plant body. Thus we understand that either through the somatic cells of plants or through the reproductive cells of animals there is *perennation* or "*no death*" in biological organisms. Biologically we could understand clearly not only *microbes* like bacteria and yeast (by binary fusion) also macroscopic living organisms do perpetuate in this universe by *passing their genomes* to their young ones by the process known as reduplication of their genetic material during reproduction that means Adam's and Eve's genomes are perpetuating (present) in this universe through their descendents. Due to *mutation* (genetic error) and *crossing over* (during meiosis, reproduction) we get variations and mixing of characters in our new ones. Thus we come across new combination (crossing over) of parental character as well as new characters (by mutation) in new species. Crossing over of meiosis during reproduction and mutation due to environmental influence are the main sources of variation.

Human beings are distinct from all other animals in their ability to acquire information and systematize it (i.e. think and act). Man has three elements-spirit, soul and body. The body encases the soul, and the soul encases the spirit. At death, body dies but the soul and the spirit live on. Almost all religions believe that life does not end with the death of the body; instead the *soul is immortal and everlasting*. Life of the body is regarded as the soul. Theologians define: "Death is a condition of extreme weakness in which all connection with the living being and God is cut off and certainly to be avoided or postponed as long as possible". Greek philosophers Plato and Plutonius believed that at death, the soul separates from the body (prison house) to attain its full and perfect state.

Through my readings of the six chosen religions I could trace two Schools of thought about soul and afterlife.

- One School of thought (very ancient) stresses on *Transmigration* of souls and "*many rebirths*" which are due to one's *Karma* (deeds) and finally *emancipation* (Hinduism, Buddhism and Jaininism).
- The other School of thought emphasises the belief in *one God* and *one life of human beings, judgment day, Heaven, Hell* and *Resurrection* (Christianity, Islam and Judaism).

In the first school of thought, though Hinduism, Buddhism and Jainism are distinct in certain beliefs, main concepts are common. "Human beings are bearers of souls *(atman)*" according to Upanishads "the soul *(atman)* persists through the process of living and dying with changing appearance of a body. As long as the soul is *entangled in the desires of the world* it continues to be born at many different levels of appearances as animals or as humans. The *rebirth* may be *for better or for worse* and is governed by a *natural law of karma*. To be born as a human being is a rare opportunity to advance towards *moksa*-release from the round of rebirth. Rebirth carries the soul through many appearances, so that rebirth has been an evil to be brought to an end.

> As a person puts on new garments,
> giving up old ones, the soul
> similarly accepts new material
> bodies, giving up the old and useless ones-Bhagavad Gita (2: 220)

The infirmity of the blind and the cripple is due to the result of his present and past karma '*As you sow, so you will reap*' has been accepted as a universal truth. Moreover, realizing the self leads to cessation of worries, fears and terror of death. Now the soul attains *mukti* or *emancipation* or *salvation*. Then the soul joins God's essence or power. Hinduism stresses the belief that the *jeevatma* after death merges with the *paramatma* just as streams flowing into the ocean lose their names and forms.

According to *Buddhism*, we can *conquer death* and *win eternal life by religious means*. The suffering of rebirth is due to our "*craving*" for sensual pleasures, for money, social position or power. A man must first of all bring some morality into his daily life by forbidding killing, stealing, indulging in sexual misconduct, lying and the use of toxicants. Further it also stresses on mental training-meditation to *withdraw one's attention from self*. Through prolonged meditation one can reach a *state of aversion* to worldly objects. Once a person inculcates the habit of aversion he reaches a state called *nirvana*. Meditation on death and aspiration of nirvana of any layman will keep him happy in this life, and he will be born in heaven. Two important doctrines of Buddhism are *avoiding doing harm* to others (ahimsa) and active *benevolence* towards them and also *fostering tenderness towards all lives*.

In addition to the transmigration of soul and rebirth concepts, *Jainism* stresses that *our passions are responsible for our karmic bondage*. Therefore controlling our passions is our ultimate victory. We should religiously cultivate self restraint, and we can improve our lives by practicing virtues like non-violence, truth and contentment. We should first of all fight our *inner enemies*. Jainism subscribes to *mutualism* or *symbiosis* and *opposes predation, parasitism, exploitation* and *violence* in any form. It also recognizes the *intrinsic interdependence* of all substances (DRAVYAS) in the universe. Such recognition guarantees protection to the environment. Any aberrations of mutualism is the root cause of all troubles, tensions, conflicts and ultimate catastrophes of individual social, national and international.

Actions of man, like his greed, carelessness, violent and irrational behaviour are the main causes for environmental degradation. Shri Gulab Singh Darda says, "The universe is neither created nor destroyed by any super power" and Jains do not believe in any superhuman entity to help us in achieving salvation.

According to Second School of Thought (Christianity, Islam and Judaism), it is certain that immediately after death; judgment takes place at which one's eternal destiny is decided. After death, when the soul leaves the body, it will be brought before the throne of God. There the "*Book of life*", reveals the good and the evil of each one, and is weighed in *God's balance of justice*. God will pass His sentence according to one's deeds. The soul alone exists in a place of eternity.

Judaism, Islam and Christianity have a common concept of reward and punishment in the afterlife. Christians stress on the *resurrection of soul and the reunion of the body* in a truly perfected state through Jesus Christ. "All men will rise from the dead in the same bodies". Christianity upholds the belief on "*Trinity of God*": "Abba the Lord Almighty, the son Jesus Christ and the Holy spirit". God has implanted consciousness in every being to prick us at the time of our ill deeds and to feel the pleasure of happiness at our good deeds. God Almighty had sent his son Jesus to help us to come out of our feelings like *lust, anger*, selfishness and wrong doings against others. Thus Christians believe in Jesus Christ as *messiah* to restore the pleasure of being with God through His (Christ's) teachings and paths to death and to resurrection.

Muslims, though they believe in God, the great Judgment, Heaven, Hell, and resurrection, feel *their actions alone* will be weighed on the balance of justice not through the Messiah like Jesus. Instead they consider Jesus as one of the prophets like Muhammad. And they address Allah as their only omnipresent God. "Every soul will be (held) in pledge for its deeds". One day every soul will come up struggling for itself and every soul will be recompensed (fully) for its action, and none will be unjustly dealt with (Al. Quran. 16:11.111).

Judaism also stresses on the judgment day, heaven, hell and resurrection. They call heaven as 'Gehenom'. In Judaism, on the great Day of Judgment (Yom Ha Din HaGadol) the soul is shown two things,

- Every decision and every thought, all the good deeds and the embarrassing things a person did in private are all shown without any embellishments (olam-HaEmet-The world of Truth). There we clearly recognize our personal strengths and shortcomings and the true purpose of life.
- How a person's life "could have been... if the *right choices had been made*, if the *opportunities were seized*, if the *potential was actualized*. Every soul feels the pain of squandered potential, but at the same time it

purifies the soul. The pain creates *regret;* which removes the barriers and enables the soul to completely connect to God. The *Jewish are awaiting for their Messiah to come and initiate the great Judgment.*

After understanding the two views of after life, one can find that the first one is emphasizing *reincarnation* where our human body is merely a *vehicle of soul,* like any other animal body. In this concept our human life is equalized to animal life. On the contrary, second view believes not only in *one life of human beings,* but also stresses on the *resurrection* where former *body will reunite with their soul* and live in the presence of God for ever. Here the *body is rewarded with eternity,* and the soul with higher perfection. Thus we could understand *resurrection is a time of reward.* And *reincarnation is a time of repairing. Resurrection is a time of reaping;* but reincarnation *is a time of sowing.*

Life is a precious gift of God. Human beings are preciously gifted with conscience and intellect which makes them distinct and superior to other animals. All human beings wish to lead a peaceful life. Peace is a condition of perfect harmony between individuals, with their environment and with God. Truth is the initiator of Peace. Peace requires respect for the God-given humanity and for the dignity of every person. Christianity preaches love for the neighbour and even goes to the extent of advocating love for the enemies.

Acceptance of the oneness of humankind is the first fundamental pre-requisite for re-organization and administration of the world as one home. 'The home of humankind' requires the abandonment of prejudice of race, colour, nation, religion, sex, civilization etc. We now need persons inspired by different religions to interact in intimate ways. A deep understanding of the concepts of death, afterlife and immortality of souls as revealed by six different religions (religious amalgamation) proves that the goal of any religion is the attainment of ultimate value or virtues. All religions can be compared to fine silk threads that constitute a network. The core idea of the fabricated network is giving up the self and rising to the level of selflessness. On removing the cocoons of desire for the world, one is able to embrace the qualities of peace, loving service, caring for the neglected and love for others. This is the ideal that a man should aspire to attain.

There is *no death to our biological body* because our genomes are carried on to the next generation through reproductive cells like sperm and eggs in human beings, animals and plants and also through the vegetative reproductive structures in plants. According to theologians also there is "*no death of soul,* and *there is afterlife*". So, it becomes evident that we are here to live with good thoughts, words and deeds. If we lead a virtuous and harmless life in this universe, *we will enjoy peace in this world and bliss in the world to come.*

ROLE OF HINDU RELIGION IN SUSTAINING ECOLOGICAL BALANCE

S.K. Khare & Jayanti Deshpande

The beliefs, practices and socio-religious institutions of the peoples known as Hindus have evolved from Vedism, the religion of the ancient Indo-European peoples who settled in India during the Second millennium B.C. Because it integrates a variety of heterogeneous elements, Hinduism constitutes whole, and because it covers the entire gamut of life, it has religious, social, economic, literary and artistic aspects. Although the various Hindu sects rely on their own set of scriptures, most also accept the sacredness of the ancient texts collectively known as the Vedas. These books are the oldest Indian documents and represent the religion of the Aryan invaders of the subcontinent over the period from 1200 to 500 B.C. Vedism was almost entirely concerned with the cult of fire sacrifice (Yajna) and the continual regeneration of the universe that resulted from it. By means of the correspondences that linked the rituals to both the macrocosmos and the microcosmos, the priest simultaneously contributed to the welfare of the transcendental order and furthered his own interests.

Out of the multitude of Vedic deities responsible for various cosmic functions, the two principal Gods of theistic Hinduism emerged as dominant at about the dawn of the Common Era. Vishnu, a God of extension and pervasiveness in the Vedas appears as Isvara or Lord in his incarnation of Krishna already in the Bhagvad Gita. Siva is described as the sole creator, preserver and destroyer of the universe as early as the Svetasvatara Upanishad (C. 300 BC) and has roots in the Vedic God Rudra. Theistic Hinduism replaced the Vedic Sacrificial cult with a form of worship or service to an image or representation of the deity called Puja and an attitude of respectful, but of ten enthusiastic devotion called bhakti. Puja consists of more or less elaborate rituals in which the deity is invoked into an image that is established either in the home or temple, honoured as a royal guest with a seat and honoured with offerings of water, garments, perfumes, flowers, food etc.

Vaisnavism, Saivism, Saktism (Devi, Durga)

Fundamental to all Hindu Sects is the belief in an eternal, infinite, all-embracing neuter principle of ultimate reality called Brahman. The pervasive force lying within all being, Brahman is conceived as the "Self" or Atman of all forms of life and many Hindu traditions portray the realization of the identity between the individual self and the cosmic principle as the final religious goal.

Mythology

Hindu mythology is rich, multifarious and inclusive. It portrays the terrible alongside the benevolent, the trivial alongside the cosmic and the grotesque alongside the sublime. The earliest source of Hindu mythology is the Vedic literature, the oldest texts of which are the four Vedas, or "Books of Knowledge": Rigveda, Yajurveda, Samaveda and Atharvaveda. Though the Vedas convey much information about the warrior life-style of the conquering Aryans, their principal focus is the fire sacrifice. These Gods most frequently invoked in the Vedas are Indra, God of wars and rain who releases the life-giving monsoons by killing the dragon Vistra, Agni, the deified sacrificial fire, and Soma, the hallucinogenic Plant from which the sacrificed drink is pressed. Most-extensive sources of Hindu mythology, however are the Puranas, encyclopedic collections of Indian legend and mythology blending Aryan and indigenous Indic elements.

Ecology and Hindu Mythology

Ecology is a science that studies interrelation i.e. mutual relationships between the living organism (biotics) and the non-living environment. It is also called Environmental Biology. Taylor described it as the Science of "all the relations of all the organisms to all the environment". There are over two million different kinds of animals and plants and ecology considers all manner of influences and interactions among them. Therefore, ecology not only includes Life Sciences but also extends Physics, Chemistry, Geography, Geology, Meteorology, Dermatology, Hydrology, Paleontology, Archaeology, Anthropology and Sociology.

Hinduism is not a religion: it is a way of life. Hindu religion teaches us to be close to animals, plants and manifestations of Nature such as rivers, seas, mountains, air etc. Since time immemorial these ways of lives have existed and maintained proximity of humans with nature and surroundings. Hindu religion is very rich culturally. There exists very vast literature in Hinduism, such as Puranas, Vedas, Upnishad, Ramayan, Mahabharat, Geeta and Jatak Stories. These scriptures contain numerous evidences of humans beings close to their surrounding. Such closeness continues even today. Various festivals and customs which are part and parcel of devout Hindus teache us to respect and love nature. Hindu religion teaches that human body is made up of five elements viz. Agni,

Vayu, Jal, Gagan and Prithvi; and when the body is put to flames these elements again go back to the respective elements present in the nature. Numerous instances find their mention in Hindu scriptures about closeness of humans and five elements.

Hindu Religion and Nature

- Prathvi is considered motherland and is worshipped as such.
- Instance of mountain pooja: Govardhan Mountain is worshipped during the festival of Diwali to commemorate the bravery of Lord Krishna who raised Govardhan Parvat on his finger to protect people from excess rains.
- Ganga Narmada and other rivers are considered sacred and worshipped. On the occasion of Poornima and Amavasya fares are organized to remind ourselves of the importance of rivers.
- Gangajal is considered very holy and is used for holy purposes.
- **Yagna**-the Agni Poojan finds mention at several places in Hindu scriptures including Ramayan.
- **Gayatri Mantra**

  Om Bhoo Om Bhuva Om Swaha, Om Maha Om Janaha Om Tapaha Om Satyam Tat Savituha Varaniyam Bhargo Devasya Dheemahi, Dhiyo Yonaha Prachodyat

  Meaning:

Bhoo	Prathvi
Bhuva	Antriksha
Swaha	Soorya
Maha	Koti Soorya
Janaha	Aakash Ganga
Tapaha	Anant Aakash Ganga
Satyam	Vishwa
Tat Savituha	Ashtva Dene Wale, Prakashman Karne Wale
Varaniyam	Sarvshrestha Parmeshwar·
Bhargo	Tum Paapnashak Ho
Devaya Dheemahi	Jo Prapt Kiya Usey Dharan Kalne Wale
Dhiyo Yonaha Prachodyat	Hamari Bhddhi Ko Prerit Karo

Recitation of Gayatri Mantra provides inner strength to an individual; it is a very unique mantra, which acknowledges the presence of nature such as the earth, the sky, the sun, the milky way and the whole universe -and they have been considered the source of energy and power and therefore worth worshipping.

Hindu Religion and Animals

- There are even instances of Naag Pooja, Naag is worshipped on the occasion of Naagpanchami.
- We compulsorily find Nandi bail (bull) at the entrance of every Shiva Temple.
- Gaj is a symbol of Lord Ganesh and Erawat Hathi is worshipped on the occasion of Mahalaxmi.
- Cow worship is a part and parcel of daily Hindu living. Demand for ban of Cow slaughter has always been made by people propagating non-violence. People like Gandhi, Vinoba Bhave have always opposed Cow slaughter.
- Cow dung is used for smearing floor and walls to make them clean and pure.
- Hindu Lords have animals as their *savaris* such as Mooshak of Lord Ganesh, Singh of Goddess Durga and Hans of Goddess Saraswati.
- Jatayu is an important part of Ramayana who fought with Ravana.
- Hanuman, the son of Wind God, is a very trusted sewak of Ram. He has played a very important part in victory of Ram over Ravana. Hanuman brings Sanjivni Booty from Himalayas when Lakshman becomes unconscious during the war. It was Hanuman who retrieved Ram and Lakshaman when they were kidnapped by Ahiravan.
- Ram has been very ably assisted by the Vanar Sena in constructing a bridge to cross over from India to Lanka.
- Jamvant, a bear was Rama's advisor.

Hindu Religion and Trees

- There is evidence of Pipal worship since Indus Valley Civilization. Lord Mahavir chose a Pipal tree for attainment of enlightenment. Ashok the Great sent Pipal saplings through his son Mahendra and daughter Sanghmitra to Ceylon from their plantation.
- Vat tree is worshipped on Amavasya in Jyeshtha month (June). Sati Savitri was able to bring back to life, her husband Satyavan by worshipping Vat tree. Since then the Vat tree is worshipped every year by Hindu married women for long life of their husbands.
- Neem tree is offered Jal by Hindus.
- Aamla tree is worshipped on Navmi after Diwali by Hindu married women for long lives of their husbands.
- Ganna or sugar cane is worshipped on Ekadashi after Diwali starting the marriage season.

Instances of Ecology as Contained in the Ramayana:

Jad chetan jag jeev jan sakal ram-maya jaani
Bandau sab ke pad kamal sada jori jug paani

All the plants and animals of the world have been equated to God.

Dev danuj nar naag khum prêt pitra gandharva
Bandau kinar rajnichar kripa karhun ab sabeyn

All Gods, Demons, Humans, Serpents, Birds, Ghosts, Eunuchs have been saluted and all have been requested to show mercy.

Aakar chari laakh churasi jaati jeev fal thal nabh bari
Siya ram-maya sab jag jaani karaun pramad jori jug paani

Four types of living beings existing in Eighty Four Lakhs of lives in water, the earth and the sky are saluted equating them to God.

Sivhi sambhu gan karhi singara. Jata mukut ahi mouri sanvara.
Kundal kankan pahire byala. Tan vibhuti pat kehri chala.

This verse relates to Shivji getting ready to be a bridegroom. His hair being decorated with snakes. He is wearing ear rings of snakes, his body is smeared with ash and in place of clothes he is wearing the skin of lion.

Yahi lalaat sandar sir ganga. Nayan teeni upbeet bhujanga
Garal kantha ur nar sir maala. Aseev vash sivdham kripala

This verse mentions that Shivji was wearing thread of Snakes and the Ganges on his forehead. Indicates love for nature.

Madhya baag saru soh suhava. Mani sopan bichitra banava
Vimal salilu sarasij baharanga. Jalkhag koojat gunjat bhranga.

This verse describes flowers and birds in Pushp Vatika.

Baagu tadagu bilouki prabhu harshey bandhu samet.
Param ramya aaramu yahu jo ramhi such det.

Indicates that Lord Ram and Shri Laxman Jee are enjoying nature.

Cahun disi chitai ponche maaligan. Lagelen dal phool mudit man
Tehi avsar sita tah aai. Girija poojan janani padhai

This refers to the Pooja of Girija Mountain by Seetaji.

Conclusion

If one follows Hindu religion strictly then respect for Ecology- Nature, Plants and Animals is automatically inculcated within us. Following Hindu Religion means trying to understand customs and traditions of the religion and also reading, understanding and interpreting Hindu scriptures such as The Ved, Upnishad, The Ramayan, The Holy Geeta and The Mahabharat.

NATURE PROCLAIMS THE GLORY OF GOD

J.V. Zahir

The word 'Nature' is defined in the Oxford Advanced Learner's Dictionary as "the whole universe and every created thing" (774). Thus, it becomes amply clear that this six letter word–'Nature' encompasses everything that has been created, whether big or small, seen or unseen, anywhere, in this vast Universe. Undoubtedly, the universe was created by God, The Almighty, The Omnipotent, The Omniscient, The Omnipresent, as per His Mighty Plan. The Holy Bible bears record of this fact, in the first book of Genesis, chapter one, we come to know how God created the Universe in six days and on the seventh day God rested "from all his work which he had made" (Gen. 2:2). For The Holy Bible says, "In the beginning God created the heaven and the earth" (Gen. 1:1) and again, "Thus the heavens and the earth were finished, and all the host of them" (Gen. 2:1). Hence, we come to know that God created the universe and named each and everything, after which, God handed over His Creation to man-Adam, and gave him the right to name it. For The Holy Bible says, "And whatsoever Adam called every living creature, that was the name thereof" (Gen. 2:19). Therefore, it is an established fact that everything seen or unseen, has been created by God, The Almighty, which undoubtedly proclaims His Glory.

Now, let us focus on God's Creation and see how it glorifies its Creator; how it teaches us to love, to share, to give; and, how it brings us closer to the reality of Life, of Beauty, and of Truth. In order to concentrate on God's Creation and to appreciate its beauty and magnificence, let us see how God created the universe in six days. To begin with the first day, The Holy Bible says, "And God said, Let there be light: and there was light. And God saw the light, that it was good: and God divided the light from the darkness. And God called the light Day, and the darkness he called Night. And the evening and the morning were the first day" (Gen. 1:3-5). These verses bring to light the fact that God created 'Light' and then God also separated it from, 'darkness' and so there was 'Day' and 'Night'. Thus God as per His Plan kept contraries as examples for man to learn and understand the deeper aspects of Life. For example, the study of the word 'light' may imply life, truth, wisdom, that which is good and the like, whereas, 'darkness' may stand for evil, death, ignorance and the like. Similarly, day and night are symbolic in nature as well. Further, let us also understand the importance of light in our lives. Let us see if man can live without light or else if life can exist without light. Certainly not.

All forms of life continue to exist as per God's Plan. For when God separated 'Day' and 'Night' it was with a definite purpose – 'Day' was meant for work and 'Night' for rest. For human it is natural to wake up early in the morning, and to sleep at night. Then again, the birds wake up mostly at dawn and before dusk they return to their nests. Then again, the trees release oxygen during the day, while they release carbondioxide at night. Thus mostly all forms of life are active during the day, while they rest at night. What does this prove? It proves that all things work and function as per The Creator's Plan who is none other than God, Almighty. And, if the universe is obeying the commands of the Lord, God, Almighty, certainly, it is proclaiming the Glory of its Creator.

About the second day of Creation the Bible says, "And God said, Let there be a firmament in the midst of the waters, and let it divide the waters from the waters. And God made the firmament, and divided the waters which were under the firmament from the waters which were above the firmament: and it was so. And God called the firmament Heaven. And the evening and the morning were the second day (Gen. 1:6-8). God created 'Heaven' which we believe to be a place where God resides, a place which is holy and pure, where truth, light, love, peace and joy prevail forever. Now the question arises. Why did God create Heaven? Certainly, God created 'Heaven' with a definite purpose. The higher purposes of making human realize that The Almighty exists. Undoubtedly, the name 'Heaven' itself reminds us of God. It makes us realize God's presence and power. The moment we look at the sky, we are reminded of God's presence, His heavenly abode. We are also made to realize and understand that there is a life beyond - a life which is eternal. Thus, we realize the reality of Life, of Beauty and of Truth. This realization itself proves that all things created by God undoubtedly glorify Him.

Moving further to the third day The Bible says: "And God said, "Let the waters under the heaven be gathered together unto one place, and let the dry land appear": and it was so.

And God called the dry land Earth; and the gathering together of the waters called he Seas: and God saw that it was good. And God said, Let the earth bring forth grass, the herb yielding seed, and the fruit tree yielding fruit after his kind, whose seed is in itself, upon the earth: and it was so. And the earth brought forth grass, and herb yielding seed, and the fruit tree yielding fruit, whose seed was in itself, after his kind: and God saw that it was good. And the evening and the morning was the third day" (Gen. 1:9-13).

If we look at the earth and the seas, and whatever is found therein, we will certainly understand that they proclaim the Glory of God. Let us look at the grass in the fields, the herbs and trees; their singular beauty reminds us of God-The Creator. For instance, mostly leaves are green but then there are unlimited shades of green. Have you ever noticed how the leaves of a particular plant and tree are different from that of another? Yes, that proves that God's Creation is a Masterpiece. Look at the innumerable species of plants, trees, herbs, shrubs, creepers, flowers, fruits, their unique beauty reminds us of The Creator. Look at the hills, mountains, valleys, plains, rivers, lakes, streams, brooks, rocks, stone, sand, all proclaim The Glory of God. Look at the earth-the things on land, the things deep down in the earth - the precious stones, rocks, coal, oils for example. Have we ever given a thought as to how the soil is replenished, how does the seed sprout under the soil, how is coal formed, how is oil formed, how are precious stones made? Yes, if man studies all this he is wonder struck. Anyone who thinks about it is astonished because everything happens as per God's Plan and all things proclaim and reveal His Power and Glory. For everything is a Masterpiece of the Master, The Creator.

Look at the huge wide 'Sea'-the beach, sand, tide, shells, corals and vegetation at the bottom of the sea-Oh! They are marvelous! Everything on land, in the water, in the sky, Glorifies God and everything works as per God's Plan. The only thing is that we need eyes to see, ears to hear, nose to smell, tongue to taste, and fingers to touch, that is to say, we need the power to proclaim and appreciate the Glory of God, His Magnificence, by using our basic senses, our mind and our heart.

On the fourth day of God's Creation God said, "Let there be lights in the firmament of the heaven to divide the day from the night; and let them be for signs, and for seasons, and for days, and years: And let them be for lights in the firmament of the heaven to give light upon the earth: and it was so. And God made two great lights; the greater light to rule the day, and the lesser light to rule the night: he made the stars also. And God set them in the firmament of the heaven to give light upon the earth, And to rule over the day and over the night, and to divide the light from the darkness: and God saw that it was good. And the evening and the morning were the fourth day" (Gen. 1:14-19).

The Sun which itself is a source of light and therefore of life, reveals the Glory of God. We know that a new day begins with the rising of the sun. As the sun rises at dawn we hear the birds singing in praise of its Creator, we hear the morning prayers being offered to God. It suggests the beginning of a new day, a new life. Have you seen the beauty of the sun at sunrise or else at sunset? Have you pondered as to why a new day begins with the break of day? The answer to all these questions is that everything created by God is unsurpassable and everything in God's Creation works only as per His Mighty, Definite Plan.

Have you not noticed and seen the beauty of the moon, the stars and the clouds? Have you not witnessed the natural phenomena of thunder and lightning? Have you not seen the rainbow? What do they proclaim? They proclaim aloud the Glory of God. For God created everything with a purpose and The Holy Bible says, "And God said, Let there be lights in the firmament of the heaven to divide the day from the night: and let them be for signs, and for seasons, and for days, and years" (Gen. 1:14). Thus, we see that everything that was created by God, The Master Craftsman, The Almighty, reveals the majesty, power and beauty of its Creator. It certainly glorifies God.

About the fifth day of creation The Holy Bible says, "And God said, Let the waters bring forth abundantly the moving creature that hath life, and fowl that may fly above the earth in the open firmament of heaven. And God created great whales, and every living creature that moveth, which the waters brought forth abundantly, after their kind, and every winged fowl after his kind: and God saw that it was good. And God blessed them, saying, 'Be fruitful, and multiply, and fill the waters in the seas, and let fowl multiply in the earth.' And the evening and the morning were the fifth day" (Gen. 1: 20-23).

Then on the sixth day God first created the creatures on the land before He made man. The Holy Bible says, "And God said, Let the earth bring forth the living creature after his kind, cattle, and creeping thing, and beast of the earth after his kind: and it was so. And God made the beast of the earth after his kind and cattle after their kind, and everything that creepeth upon the earth after his kind: and God saw that it was good" (Gen. 1:24-25).

Thus, we come to know that God created the creatures of the water and the fowl of the air on the fifth day, while He created the creatures of the land on the sixth day. It is a fact that every animal, insect, bird that we can find on this planet 'Earth' is a masterpiece. We also know that no man is able to see all the species of animals, insects or birds created by God because they are innumerable. But whatever form of life we see around us it makes us realize the Greatness of God. We find these creatures flourishing in their own environment as they were blessed by God.

Moreover, we need to learn from them for the Word of God says: "Take no thought for your life, what ye shall eat: neither for the body, what ye shall put on. The life is more than meat, and the body is more than raiment. Consider the ravens: for they neither sow nor reap; which neither has storehouse nor barn; and God feedeth them: how much more are ye better than the fowls? ... If then God so clothe the grass, which is to-day in the field, and tomorrow is cast into the oven; how much more will he clothe you ... And seek not ye what ye shall eat, or what ye shall drink, neither be ye of doubtful mind ... But rather seek ye the kingdom of God; and all those things shall be added unto you" (Luke 12:22-31).

Further, these creatures whether big or small are examples for man in different respects. For instance, the song of the birds early in the morning, the faithfulness of the dog, the meekness of a lamb. The Holy Bible also says, "Be ye therefore wise as serpents, and harmless as doves" (Matt. 10:16). Thus, it becomes amply clear that God made all things and it is our duty to understand all forms of life as well as to look after them. For everything was created by God with a definite purpose and all things created by Him reveal His Glory.

Focusing again on the sixth day of God's Creation, after God had created the living creatures of the earth, we come to know that God created man. The Holy Bible says, "So God created man in his own image, in the image of God created he him" (Gen. 1:27). Then again, The Bible says, "And the Lord God formed man of the dust of the ground, and breathed into his nostrils the breath of life; and man became a living soul" (Gen. 2:7). Undoubtedly, God created man in his own 'likeness' but out of dust. The Word of God confirms this fact. For it states: "Dust thou art, and unto dust shalt thou return" (Gen. 3:19). Even science bears witness to the fact that once a person dies he is reduced to dust though it takes time. Then again, we know that when a person is critically ill, even the doctors give up hope and say, "Now everything is in God's Hands." What does this prove? Certainly, it proves the fact that God is The Creator of all things and everything happens as per His Will and this fact itself Glorifies God. For instance, let us take 'Birth' and 'Death'. Definitely, everyone will agree to the fact that 'Birth' and 'Death' are totally in God's Hands. Further, God has blessed man with innumerable talents and gifts, the power to think, to study and to analyze all things created by Him. This proves the fact that God is Almighty as He has made every human being unique which reveals His Glory and Majesty.

But what has man done to God's Creation? He has basically misused his talents, abilities and power to spoil and destroy God's Creation. That is why, today, we are faced with serious problems of all kinds such as ecological, social, political and the like. Today, man has forgotten the fact that God created him in his own 'likeness' - a spiritual being - with a definite purpose. Man has forgotten the fact that he was created with a higher purpose to have a personal relationship with his Creator. He has also forgotten his responsibility of looking after all the living creatures whom God had 'delivered' into his hands. What does this prove? It proves the fact that man who is not sure of his existence, who does not know what is going to happen in the immediate future, has drifted away from God. Thus, we are faced with the harsh reality of life that man is the only one in God's entire Creation who fails on his part to proclaim the Glory of God. This is so because he fails to seek the Kingdom of God and even if he tries to do so he fails to do it with his whole heart, mind and soul with Truth. However, at the same time there are very few people in the world who proclaim God's Glory by their words and deeds, and who worship Him "in spirit and in truth." For The Bible says, "God is a Spirit: and they that worship him must worship him in spirit and in truth" (John 4:24).

Thus, 'Nature' created by God, undoubtedly, proclaims the glory of its Creator. Every aspect of nature, everything that exists, praises God in its own way. Nature which is a constant Teacher reminds us at every moment that God, The Creator is Almighty. Undoubtedly, it teaches us to love, to share, to give, to comfort, to provide solace and peace of mind. Indeed, it brings us closer to the reality of Life, of Beauty and of Truth. As John Keats rightly opined in his famous poem "Ode on A Grecian Urn": 'Beauty is truth, truth beauty' - that is all and Ye know on earth, and all ye need to know" (359). 'Nature' ie., God's Creation which comprises of "the whole universe" (774) Glorifies Him, who is Almighty. Everything that was created by God in the beginning continues to Proclaim God's Glory unceasingly and it will continue to do so as per God's Plan for God Himself says, "I am Alpha and Omega, the beginning and the end, the first and the last" (Rev. 22:13). And till the time the universe will continue to exist as per God's Plan, till that time the universe will proclaim His Glory and Might, for God is Omnipotent, Omniscient and Omnipresent.

References

The Holy Bible. King James Version. New York: American Bible Society, 1982.

Fifteen Poets. London: OUP, Ely House, 1968.

"Nature". *Oxford Advanced Learners's Dictionary of Current English.* (ed.) Jonathan Crowther, et al. Bombay: OUP, 1997.

